Agriculture Issues and Policies Series

ENVIRONMENTAL SERVICES AND AGRICULTURE

AGRICULTURE ISSUES AND POLICIES SERIES

Agriculture Issues & Policies, Volume I
Alexander Berk
2001. ISBN 1-56072-947-3

Hired Farmworkers: Profile and Labor Issues
Rea S. Berube
2009. ISBN 978-1-60741-232-8

Effects of Liberalizing World Agricultural Trade
Henrik J. Ehrstrom
2009 ISBN: 978-1-60741-198-7

Environmental Services and Agriculture
Karl T. Poston
2009 ISBN: 978-1-60741-053-9

Agriculture Issues and Policies Series

ENVIRONMENTAL SERVICES AND AGRICULTURE

KARL T. POSTON
EDITOR

Nova Science Publishers, Inc.
New York

Copyright © 2009 by Nova Science Publishers, Inc.

All rights reserved. No part of this book may be reproduced, stored in a retrieval system or transmitted in any form or by any means: electronic, electrostatic, magnetic, tape, mechanical photocopying, recording or otherwise without the written permission of the Publisher.

For permission to use material from this book please contact us:
Telephone 631-231-7269; Fax 631-231-8175
Web Site: http://www.novapublishers.com

NOTICE TO THE READER

The Publisher has taken reasonable care in the preparation of this book, but makes no expressed or implied warranty of any kind and assumes no responsibility for any errors or omissions. No liability is assumed for incidental or consequential damages in connection with or arising out of information contained in this book. The Publisher shall not be liable for any special, consequential, or exemplary damages resulting, in whole or in part, from the readers' use of, or reliance upon, this material.

Independent verification should be sought for any data, advice or recommendations contained in this book. In addition, no responsibility is assumed by the publisher for any injury and/or damage to persons or property arising from any methods, products, instructions, ideas or otherwise contained in this publication.

This publication is designed to provide accurate and authoritative information with regard to the subject matter covered herein. It is sold with the clear understanding that the Publisher is not engaged in rendering legal or any other professional services. If legal or any other expert assistance is required, the services of a competent person should be sought. FROM A DECLARATION OF PARTICIPANTS JOINTLY ADOPTED BY A COMMITTEE OF THE AMERICAN BAR ASSOCIATION AND A COMMITTEE OF PUBLISHERS.

LIBRARY OF CONGRESS CATALOGING-IN-PUBLICATION DATA
Available upon request
ISBN: 978-1-60741-053-9

Published by Nova Science Publishers, Inc. ✦ New York

CONTENTS

Preface		vii
Summary		ix
Chapter 1	Introduction	1
Chapter 2	Environmental Services from Agriculture	7
Chapter 3	Market Basics	15
Chapter 4	What Can We Learn from Current Markets?	23
Chapter 5	Lessons Learned and Potential Roles for Government	71
References		83
Appendix: Predicting the Location of New Mitigation Banks		93
Index		97

PREFACE*

U.S. farmers and ranchers produce a wide variety of commodities for food, fuel, and fi ber in response to market signals. Farms also contain signifi cant amounts of natural resources that can provide a host of environmental services, including cleaner air and water, flood control, and improved wildlife habitat. Environmental services are often valued by society, but because they are a public good—that is, people can obtain them without paying for them—farmers and ranchers may not benefi t financially from producing them. As a result, farmers and ranchers underprovide these services. This report explores the use of market mechanisms, such as emissions trading and eco-labels, to increase private investment in environmental stewardship. Such investments could complement or even replace public investments in traditional conservation programs. The report also defi nes roles for government in the creation and function of markets for environmental services.

Keywords: *Eco-labeling, environmental service, emissions trading, market, public good, supply and demand, transaction cost*

---

* This is an edited, excerpted and augmented edition of a United States Department of Agriculture publication.

Summary

U.S. farmers and ranchers produce a wide variety of commodities for food, fuel, and fiber in response to market signals. Farms also contain significant amounts of natural resources that can provide a host of environmental services, including cleaner air and water, flood control, and improved wildlife habitat. Environmental services are often valued by society, but because they are a public good—that is, people can obtain them without paying for them—farmers and ranchers may not benefit financially from producing them. As a result, farmers and ranchers underprovide these services.

What Is the Issue?

Farmers can provide environmental services by adopting conservation or production practices that improve the environment. Farmers often produce these services unintentionally, however, by maintaining grasslands, wetlands, or forests rather than converting them to cropland or by adopting practices that increase net returns but also improve environmental performance. Although society values these services, because of the services' public-goods nature, farmers usually cannot benefit financially by intentionally producing them. As a result, there are no naturally occurring markets for environmental services. If environmental services could be sold like other commodities, farmers would likely invest more to maintain wildlife habitat, woodlots, and wetlands. The U.S. Department of Agriculture (USDA) has expressed great interest in the creation of markets to provide environmental quality and other environmental services. Such markets would supplement existing conservation programs and provide an additional source of income for farmers.

What Did the Study Find?

Markets for environmental services may fail to form or function properly for several reasons.

- The public-goods nature of most environmental services is the primary reason that markets for them do not naturally develop. In addition, environmental services, such as improved water quality and wildlife preservation, are unintended consequences of the primary production activities on the farm. These characteristics can limit potential suppliers' ability to benefi t financially from providing environmental services.
- Uncertainty about the quantity and quality of services a farmer can produce is a common problem that often hinders market function. Environmental services are often diffi cult to observe, such as the nutrientfi ltering capacity of wetlands or the sequestration (storing) of greenhouse gases from adopting conservation tillage. Farmers are reluctant to adopt management practices if potential returns are uncertain. Uncertain quality can also deter potential buyers from purchasing environmental services from farms.
- Environmental services are associated with the land and are not transportable to central markets. The costs of bringing buyers and sellers together may hinder the development of markets.
- Government conservation programs and markets for environmental services sometimes have common objectives and outcomes and may end up competing for the same land, the natural capital in the production of environmental services. Such competition could hinder the development of markets by driving up costs.

The consequence of these limitations is that markets for environmental services are rare. Even though public demand for environmental services is strong, farmers are unable to benefi t financially by providing them.

Barriers to market development and function can be overcome in a number of ways.

- In some cases, regulation can be used to create a private good, and the demand for that good, that is closely related to an environmental service. For example, the Federal Government places caps on pollutant discharges from regulated firms and issues discharge allowances to each firm, specifying how much pollution the firm can legally discharge. A firm may be

able to discharge more pollution than its original allocation by purchasing allowances from other firms that have cut their own pollution discharges below their own allowances or from unregulated sources of pollution, such as agriculture. This transaction is known as a trade. Discharge allowances, therefore, have characteristics of a private good. Farmers are often able to provide discharge reductions at a lower unit cost than industry can and to profi t from the exchange.
- Uncertainty over the performance of agricultural management practices for the production of environmental services can be reduced through education and research. USDA and State efforts can play an important role in both areas. Research at the Agricultural Research Service, USDA, and the Conservation Effects Assessment Project at the Natural Resources Conservation Service, USDA, are quantifying the performance of management practices in different settings, and State extension services can convey this information to farmers. In addition, validation and certifi cation services can bolster consumer confi dence that, when they purchase environmental services, they are getting the service for which they paid. USDA has played an important certifi cation role in the organic market.
- Improved market design can reduce the search and bargaining costs of bringing buyers and sellers together. Government or other entities can play the role of an aggregator or clearinghouse in a market, making it easier for geographically dispersed market participants to find each other, thereby reducing bargaining costs.

Coordinating conservation programs and environmental service markets can enhance the performance of both. Targeting conservation programs to producers who need to meet minimum performance standards to enter a market would likely increase the number of farmers willing to participate. Identifying program rules that prevent farmers from selling environmental services for which they have not received a government payment would also increase farmer interest in entering environmental service markets.

Creating markets for environmental services is not always possible or advisable. Transactions costs associated with reducing uncertainty may be greater than the benefi ts of creating a market. The public-goods nature of environmental services may also prevent markets from developing, despite research and education. Even though people may be willing to pay for environmental services, the ability to acquire these services without paying for them reduces the incentive for farmers to provide them. In these cases, regulation or direct financial assistance through government programs may be the most cost-effective options.

How Was the Study Conducted?

The study used an extensive literature review and five case studies to explore important economic issues affecting the development of markets for environmental services. Because working markets for environmental services are rare, we used the literature to provide the reasons that markets are not developing and to provide insight into the role government might play in helping markets to form and to function.

We present case studies for environmental services for which attempts have been made to develop markets. These markets are as follows:

- Water quality trading—Firms with high pollution-control costs purchase pollution reductions from another source at lower cost.
- Carbon emissions trading—Same as water quality trading.
- Wetland mitigation—Loss in wetland services is offset by an improved wetland with similar services.
- Fee hunting—Hunters pay for access to land in order to hunt.
- Eco-labeling—Labels tout goods made in a way that avoids harming the environment.

These case studies provide a more detailed look at the issues surrounding markets for environmental services, as well as the steps that were taken to overcome market impediments. The findings of the case studies are used to identify some specific actions governments could take to support the creation and function of markets for environmental services. This report provides context for the actions USDA has recently taken to support markets for environmental services and for the Department's response to the Food, Conservation, and Energy Act of 2008.

Chapter 1

INTRODUCTION

Farmers and ranchers produce a wide variety of agricultural commodities, which are sold in well-established markets. Farms and ranches can also produce a variety of environmental services that are often unintended consequences of production practices or land use decisions. Some examples are air and water, flood mitigation, drought mitigation, and wildlife. Even when unintended, these services provide benefits to people. Agricultural producers' actions can increase or decrease the provision of environmental services. Understanding how agricultural producers make their production and land management decisions is critical in designing strategies for enhancing those environmental services that people value.

Well-functioning commodity and input markets use prices to signal farmers and ranchers what to produce with their land and how to allocate resources most efficiently to maximize profits. In contrast, for a variety of reasons, markets for environmental services have generally not developed. As a result, producers' responses to market signals lead them to produce agricultural commodities rather than environmental services. Environmental services therefore may be underprovided from society's point of view.

Yet, with growing population and incomes, society increasingly values the environmental services agriculture can produce (Antle, 1999). Since markets typically undersupply environmental services, Federal, State, and local governments have developed a range of approaches for increasing their production (table 1.1). Most rely on policy tools, such as financial and technical assistance, regulation, and education. Although these approaches may be relatively simple to implement, basic economic principles suggest that they cannot allocate resources as efficiently as working markets, assuming such markets can exist.

Table 1.1. Matrix of Federal agricultural conservation/environmental policy instruments and problems

	Participation							
	Involuntary		Voluntary					Facilitative
	Regulation	Conservation compliance	Taxes	Land retirement	Cost sharing	Incentive payments	Markets (Trading/ offsets/ labeling)[1]	Education/ technical assistance
Problem:	Instrument							
Erosion: Soil productivity		Sodbuster/ compliance (1985)		Soil Bank (1956-60) CRP (1985)	ACP (1936-96) EQIP (1996)	CSP (2002) EQIP (1996)		CTA (1936) CEP (1914)
Erosion: sedimentation	CZARA (1990)	Sodbuster/ compliance (1990)		CRP (1990)	ACP (1936-96) EQIP (1996)	WQIP (1990-96) EQIP (1996) CSP (2002)		CTA (1936) CEP (1914)
Erosion: airborne dust	Clean Air Act	Sodbuster/ compliance (1990)		CRP (1996)	ACP (1936-96) EQIP (1996)	WQIP (1990-96) EQIP (1996) CSP (2002)		CTA (1936) CEP (1914)
Wetlands	CWA Section 404 (1972)	Swampbuster (1985)		Water Bank (1970-95) CRP (1988) WRP (1990) EWRP (1993)			Mitigation banking (1995)	CTPA (1936) CE (1914)

	Participation							
	Involuntary		Voluntary					Facilitative
	Regulation	Conservation compliance	Taxes	Land retirement	Cost sharing	Incentive payments	Markets (Trading/ offsets/ labeling)[1]	Education/ technical assistance
Problem:	**Instrument**							
Water quality: nutrients	CWA Section 402 (2003)			CRP (1996)	EQIP (1996)	WQIP (1990-96) EQIP (1996) CSP (2002)	CWA (1990)	CTA (1936) CEP (1914)
Water quality: pesticides	FIFRA (1947) CZARA (1990)			CRP (1996)	EQIP (1996)	WQIP (1990-96) EQIP (1996) CSP (2002)		CTA (1936) CEP (1914)
Wildlife habitat	ESA (1973)			CRP (1996) GRP (2002)	WHIP (1996)	EQIP (1996) CSP (2002)	Conserva-tion banking (2003) Eco-labeling	CTA (1936) CEP (1914)

Acronyms:

ACP—Agricultural Conservation Program, CEP—Cooperative Extension, CRP—Conservation Reserve Program, CSP—Conservation Security Program, CTA—Conservation Technical Assistance, CWA—Clean Water Act, CZARA—Coastal Zone Act Reauthorization Amendments, EQIP—Environmental Quality Incentives Program, ESA—Endangered Species Act, EWRP—Emergency Wetland Reserve Program, FIFRA—Federal Insecticide, Fungicide, and Rodenticide Act, GRP—Grassland Reserve Program, WHIP—Wildlife Habitat Incentives Program, WQIP—Water Quality Improvement Program, WRP—Wetland Reserve Program.

Note: Year denotes first year Federal program authorized.

[1] Trading and offsets rely on regulatory measures to create a market. However, agriculture's participation is currently voluntary.

The U.S. Department of Agriculture (USDA) and other groups have expressed great interest in the use of market-based policy instruments as a more effi cient way of providing environmental quality and other environmental services. In 2006, USDA outlined its role in "market-based environmental stewardship." USDA is seeking to broaden the use of markets for environmental goods and services to "...encourage competition, spur innovation, and achieve environmental benefi ts..." (USDA, Natural Resources Conservation Service, 2006b). Some of the approaches that can be used to promote markets include credit trading, mitigation banking, and ecolabeling. To emphasize USDA's growing role, the Food, Conservation, and Energy Act of 2008 includes a provision directing USDA to facilitate the participation of farmers, ranchers, and forest landowners in environmental services markets. The U.S. Environmental Protection Agency (EPA) is promoting emissions trading as a way of reducing the cost of meeting air and water quality goals. The Organisation for Economic Co-Operation and Development is also promoting the use of market mechanisms for the provision of environmental services (Organisation for Economic Co-Operation and Development, 2005).

Creating markets for environmental services is no simple task. A key measure of a well-functioning market is how well it facilitates interaction between consumers and producers, which involves much more than simply the sale of environmental services.[1] A sustainable market should be based on more-or-less direct interaction between demanders and suppliers without constant government intervention when unanticipated changes occur.

The purpose of this report is to explore the conditions under which markets for environmental services from agriculture might arise and when and how government intervention might help environmental service markets succeed. This report presents an extensive review of the types of environmental services farmers can produce, what is required for a market to form, and the problems these markets might face in functioning smoothly. We consider potential roles for government in creating and supporting a market, with a focus on reducing transaction costs.

The report also assesses the potential supply of environmental services to provide a perspective on the potential scale of such markets. By providing a

---

[1] Payments to agricultural producers for the production of environmental services are fairly common. USDA currently supports the production of environmental services through conservation programs, such as the Environmental Quality Incentive Program, Wildlife Habitat Incentive Program, Conservation Reserve Program, and Wetland Reserve Program. Land trusts, such as the Nature Conservancy and Ducks Unlimited, purchase land or easement to land in order to protect the flow of environmental services, primarily wildlife or biodiversity.

clearer, stronger, more systematic motivation for government intervention in the development of environmental service markets, this report provides insight on ways in which government actions might link the public's demand for environmental services to agriculture's supply of these services and on conditions under which the formation of markets, despite government actions, is impracticable.

Chapter 2

ENVIRONMENTAL SERVICES FROM AGRICULTURE

Environmental services from agriculture are a subset of ecosystem services from agriculture. Ecosystem services are defi ned by the United Nations' Millennium Ecosystem Assessment as the "benefi ts people obtain from ecosystems" (Millennium Ecosystem Assessment, 2003, p. 3). These include a wide range of provisioning, regulating, cultural, and supporting services. Both unmanaged and managed ecosystems (such as agricultural lands) can provide these services.

Farmers and ranchers constitute the largest group of natural resource managers in the world (Food and Agriculture Organization of the United Nations, 2007). Farms exist to produce food, fuel, and fiber and to sell them to consumers. However, farms also produce many other ecosystem services as externalities, in that they are unintended consequences of the primary production activities on the farm and those who are affected cannot infl uence their production. Farms can produce externalities as part of the production process (generally negative externalities, such as nutrient runoff or air pollution) or land use decisions (positive externalities, such as wildlife, wetland services, and water quality from farmland not planted to crops).

In this report, environmental services refer to positive externalities that result from stewardship on the farm (table 2.1). These externalities could include improved water quality from changes in crop management, carbon sequestration from converting cropland to forests, wetland services from preserving a wetland, and enhanced wildlife habitat by providing adequate food, cover, and nesting habitat.

Table 2.1. Some environmental services and farm management options

Environmental service	Farm-level management option
Carbon sequestration in soils	Manage soil organic matter
Carbon sequestration in perennial plants	Convert cropland to grassland or forest
Methane emission reduction	Capture and destroy methane from animal waste storage structures
Water quality maintenance	Reduce agrichemical use, establish vegetative buffers, improve nutrient management
Erosion and sediment control	Manage soil conservation and runoff, increase soil cover
Flood control	Create diversions, wetlands, storage ponds
Salinization and water table regulation	Grow trees, manage water
Wildlife	Protect breeding areas and wild food sources, improve timing of cultivation, increase crop species/varietal diversity, reduce use of toxic chemicals

Source: Food and Agriculture Organization of the United Nations, 2007.

Natural capital possesses the capacity of giving rise to the flow of environmental services (Boyd and Banzhaf, 2006; Costanza et al., 1997; Elkins, 2003). The natural capital that agricultural producers control is the land, water, air, and genetic resources on their farms. How these resources are managed affects the type and level of environmental services that can be produced.

Agriculture controls a large amount of natural capital in the United States. In 2002, private farms accounted for 41 percent of all U.S. land, including 434 million acres of cropland, 395 million acres of pasture and range, and 76 million acres of forest and woodland (USDA, National Agricultural Statistics Service, 2004). This capital can provide a host of environmental services, including water quality, air quality, flood control, wildlife, and carbon sequestration. These services can be consumed directly or combined by consumers with other goods to create final goods, such as sightseeing, fishing, wildlife viewing, or hunting. In this report, we focus on the provision of water quality, greenhouse gas reduction, wildlife, and wetland services. Markets have been developed for providing these services, and these are the ones specifi cally mentioned in the USDA policy on markets for environmental services (USDA, Natural Resources Conservation Service, 2006b).

WATER QUALITY

The potential for agriculture to supply water quality improvement is defined largely by the significant negative impact that agriculture has historically had on water quality. Current production practices and inputs used by agriculture can result in a number of pollutants—including sediment, nutrients, pathogens, pesticides, and salts—entering water systems. Pollution from agriculture is generally exempt from regulations under the Clean Water Act, so agricultural producers have little incentive to address these largely offsite impacts.

Although no comprehensive national study of agriculture and water quality has been conducted, the magnitude of the impacts can be inferred from several water quality assessments. EPA's 2000 Water Quality Inventory reports that agriculture is the leading source of pollution in 48 percent of river miles, 41 percent of lake acres (excluding the Great Lakes), and 18 percent of estuarine waters that are impaired. The inventory shows these bodies of water to be water-quality impaired in that they do not support designated uses, such as swimming and aquatic life (U.S. EPA, 2002). The findings mean that agriculture is the leading source of impairment in the Nation's rivers and lakes and a major source of impairment in estuaries.

Agricultural producers can improve water quality by reducing the discharge of nutrients, pesticides, sediment, and other agricultural pollutants to water resources. The Natural Resources Conservation Service's (NRCS) technical field guide lists over 300 management practices that can improve water quality (USDA, NRCS, 2007b). These practices include conservation tillage, nutrient management, strip cropping, irrigation management, pesticide management, manure storage structures, vegetative buffer strips, fencing, and livestock watering facilities. Farmers can also retire cropland in sensitive areas and improve or restore wetlands to filter sediment and nutrients.

AIR QUALITY

Agricultural production releases a wide variety of material into the air. Field operations produce windblown soil, nitrogen gases, and pesticides. Animal operations release hydrogen sulfide, ammonia, methane, volatile organic compounds, and odors. Internal combustion engines in field equipment and irrigation pumps and field burning produce fine particulates and nitrogen oxides. These pollutants may affect people's health, reduce visibility, and contribute to global warming or may simply be a nuisance. Agriculture can improve air quality by reducing the release of these materials through changes in soil, water, chemical, and manure management.

Greenhouse gases have been of particular recent interest due to their role in global climate change. Agriculture is both a source and a sink (storage in soil and in biomass) of greenhouse gases. It is a relatively small source of greenhouse gas emissions, accounting for about 8 percent of all U.S. greenhouse gas (GHG) emissions in 2005 (USDA, Offi ce of the Chief Economist, 2007). The most important GHG emissions from agriculture are nitrous oxide (N2O) and methane (CH4). Agricultural soil management (60 percent), enteric fermentation (25 percent), manure management (13 percent), rice cultivation (2 percent), and agricultural residue burning (less than 1 percent) are the sources of agricultural GHG emissions.

Agriculture can sequester (store) carbon in soils and biomass, thus offsetting GHG emissions. Carbon entering the soil is stored primarily as soil organic matter. Agricultural soils sequestered an estimated 12.4 million metric tons carbon equivalent in 2004, less than 1 percent of U.S. emissions (U.S. EPA, Offi ce of Atmospheric Program, 2006). Studies indicate that it may be technically possible to sequester an additional 89-318 million metric tons of carbon annually on U.S. croplands and grazing lands through various management practices, such as conservation tillage, crop rotations, and fertilizer management, or up to 16 percent of 2004 emissions (Lewandrowski et al., 2004). Shifting cropland to grasslands or forest could increase sequestration even more.

WILDLIFE

U.S. agriculture is in a unique position with respect to the Nation's wildlife resources. The historic development of U.S. agriculture required the development of large amounts of native grasslands, wetlands, and forests for agricultural purposes. Management of the Nation's farms and ranches can play a major role in protecting and enhancing its wildlife. Because of the dominance of private land ownership in the United States, Federal and State governments cannot exercise effective responsibility for wildlife management without productive collaboration with private land managers (Benson, 2001b; Conover, 1998).

The quality of wildlife resources is a function of the amount, quality, and diversity of habitat. Grasslands and wetlands are two common types of habitat that can be protected, restored, or improved through conservation on agricultural lands.

Grassland Habitat

Grasslands constitute the largest land cover on America's private lands. These lands provide biodiversity of plant and animal populations and play a key role in environmental quality. Grasslands also improve the aesthetic character of the landscape, provide scenic vistas, open spaces, and recreational opportunities, and protect soil from water and wind erosion.

Large expanses of grassland acreage are annually threatened by conversion to other land uses, such as cropland and urban development. About half of all U.S. grasslands have been lost since settlement, much due to conversion to agricultural uses (Conner et al., 2001).

Wetland Habitat

Wetlands are complex ecosystems that provide many ecological functions valued by society. They take many forms, including prairie potholes, bottomland hardwood swamps, coastal salt marshes, and playa wetlands. Wetlands are known to be the most biologically productive ecosystems in temperate regions. More than a third of threatened and endangered species in the United States live only in wetlands, and nearly half use wetlands at some point in their lives (U.S. EPA, Office of Water, 1995a). Most freshwater fish depend on wetlands at some stage of their lives. Many bird species depend on wetlands for either resting places during migration, nesting or feeding grounds, or cover from predators. Wetlands are also critical habitat for many amphibians and fur-bearing mammals. Besides supporting wildlife, wetlands also supply water pollution control, flood control, water supply protection, and recreation.

When the country was first settled, there were 221-224 million acres of wetlands in the continental United States (Heimlich et al., 1998). Since then, about half have been drained and converted to other uses, nearly 85 percent for agricultural uses. Currently, there are about 111 million acres of wetlands on non-Federal lands (USDA, NRCS, 2004b). About 15 percent are on agricultural lands (cropland, pastureland, and rangeland).

DEMAND FOR ENVIRONMENTAL SERVICES

The existence of a market for an environmental service requires that potential consumers are willing to pay a price for those services. Numerous studies have found that people are in fact willing to pay for environmental services from agriculture (Environmental Valuation Reference Inventory, 2007). These findings do not mean a market should exist, but they are a prerequisite for a market to exist.

Another indication that demand for environmental services exists is that State and Federal governments have developed many programs to supply them, implicitly refl ecting public demand. Conservation programs, such as the Conservation Reserve Program, Wetland Reserve Program, Environmental Quality Incentives Program, and Farm and Ranch Protection Program, provide financial and technical incentives to agricultural producers to retire land, adopt management practices that protect and enhance environmental quality, or preserve farmland. In recent years, USDA has spent over $4.5 billion per year on such programs (USDA, Economic Research Service, 2007a).

Environmental regulations are also used to ensure that environmental services are provided. Regulations in the Clean Water Act; Clean Air Act; Federal Insecticide, Fungicide, and Rodenticide Act; and Endangered Species Act keep harmful chemicals from water and air, prevent wetland loss, and protect habitat for endangered species. These and other regulations have been created because the public demands that environmental services be protected. Agriculture, however, is often exempt from these regulations, which leaves other mechanisms, such as financial assistance, to provide incentives for agriculture to increase its production of environmental services.

Demand for environmental services can also be expressed through private actions, such as the purchase of conservation easements by land trusts. Land trusts are one alternative mechanism by which individuals can choose to act privately to address the failure of both governments and private markets to provide environmental services (Sundberg, 2006). They preserve and increase environmental services, based on the perception of their members' interests, by obtaining fee title or conservation easements of land they want to protect. The United States had over 1,600 land trusts, protecting open space and habitat on over 37 million acres of land, in 2005 (Land Trust Alliance, 2006). About 1.2 million acres per year are added to the rolls of privately conserved land.

SUMMARY

Agriculture controls natural capital that can provide environmental services. There is much evidence that people value these services, yet there is longstanding concern over their continued loss. Government programs and nongovernment efforts have been developed to motivate agricultural producers to provide environmental services. These efforts raise the question of why landowners do not market and sell these services as they do agricultural commodities, thereby attaining an additional stream of income. The next chapter reviews the function of markets and the reasons that markets fail to develop.

Chapter 3

MARKET BASICS

Markets are institutions through which potential buyers and sellers of goods and services deal with each other in the process of exchange. In a perfect world of competitive markets, resources move to their highest valued use (see box, "Value of Markets"). With market failure—that is, when markets do not operate properly—resources are not allocated to their highest valued use. Addressing market failure is one of the roles of government.

MARKETS FOR ENVIRONMENTAL SERVICES

Few well-functioning markets have developed for environmental services, even though evidence is strong that consumers are willing to pay for them (see chapter 2). The lack of markets has important consequences in the allocation of resources on farms. Without well-defined markets for environmental services, landowners are not rewarded financially for supplying them. For example, without a market for environmental services, a farmer with native vegetation on his or her land has no economic incentive to preserve the cover and the environmental services it provides. The farmer's land-use decision will be based on the potential return from agricultural commodities. If the value that society places on environmental services could be captured by the farmer, he or she would more likely keep a larger fraction of his or her land in a natural state.

Keep in mind that agricultural producers' motivations are more complex than simply profit maximization. Most agricultural producers value environmental services and may sacrifice some potential income to enjoy them on their farms. Without markets, however, agricultural producers' provision of environmental services is based on their own personal preferences, rather than the value society places on them. The result is likely to be an underprovision of those services.

WHY DO MARKETS FAIL?

Before exploring how markets for environmental services might be created, it is important to understand why markets fail. Markets for environmental services rarely exist because one or more of the following factors apply (Murtough, Aretino, and Matysek, 2002; Ruhl, Kraft, and Lant, 2007):

- Public good characteristics.
- Market burdens, such as large transaction costs and uncertainty.
- Institutional barriers.

Public Goods

Because environmental services are the product of complex ecosystem processes and delivered through a variety of landscape settings, they nearly always take on characteristics of public goods; they are nonexcludable and nonrival. With a private good, a producer can prevent someone who has not paid for it from obtaining it; it is excludable. For a public good, a provider cannot exclude someone who has not paid a price from obtaining it. For example, a farmer contemplating the sale of improved water quality by establishing vegetative buffers on his or her farm cannot exclude downstream users from benefi ting; the downstream users are "free riders." In this situation, the farmer does not have an economic incentive to provide the good.

Furthermore, when a good is nonrival—that is, exclusive ownership is not possible—a buyer's purchase of a good will also benefi t other individuals. Thus, the value to society of the good (say, improved water quality) is the sum of everyone's enjoyment. However, when individuals consider how much they will pay, they will not consider this sum; instead, they consider only their own personal values. Thus, even if a willing seller existed, the net price the producer could receive would be too low; it would refl ect one individual' s value rather than the sum of the values of all individuals.

Markets are driven by individuals and firms striving to maximize their own wellbeing. Relative prices determined by the interaction of supply and demand satisfy the necessary conditions for maximizing social welfare. The producer combines price information from product markets with price information from input markets to determine how much to produce and how many inputs to purchase. The market supply curve for the product represents the production decisions made by all producers over a range of prices. The consumer participates in various product markets based on prices, income, and personal preferences. The market demand curve for a product represents the purchasing decisions made by all consumers over a range of prices.

If the market functions properly, factors of production move to those uses where they earn the highest return; resources are used most efficiently, and both producers and consumers enjoy maximum benefits from production and consumption.

Prices in a perfectly operating market tell participants how valuable one good or input is relative to another, making them the most important piece of information driving decisions about production and consumption. However, markets rarely operate perfectly. Various factors can affect the interplay of supply and demand so that prices no longer convey the true values of goods and services. A market for a good may also fail to form entirely. Under these conditions, factors of production do not move to those uses where they have the greatest value; resources are misallocated, and overall social welfare is lower than if markets operated perfectly.

The figure depicts the production possibility frontier (PPF) for a farm, or the marginal tradeoff between production of a commodity and an environmental service. The shape of the curve is a function of the farm's resource base and technology set and the farmer's management skills. The mix of commodities and environmental services provided by the farmer depends on the prices received for each. If no market exists for the environmental services and only the commodity has a price, then the farmer maximizes income by producing at point A; no environmental services are produced. Alternatively, if a market for environmental services could be created, then a price for that service would exist. The farmer would maximize net returns by producing at a point such as B, where the slope of the PPF equals the ratio of prices. Fewer commodities and more environmental services are produced.

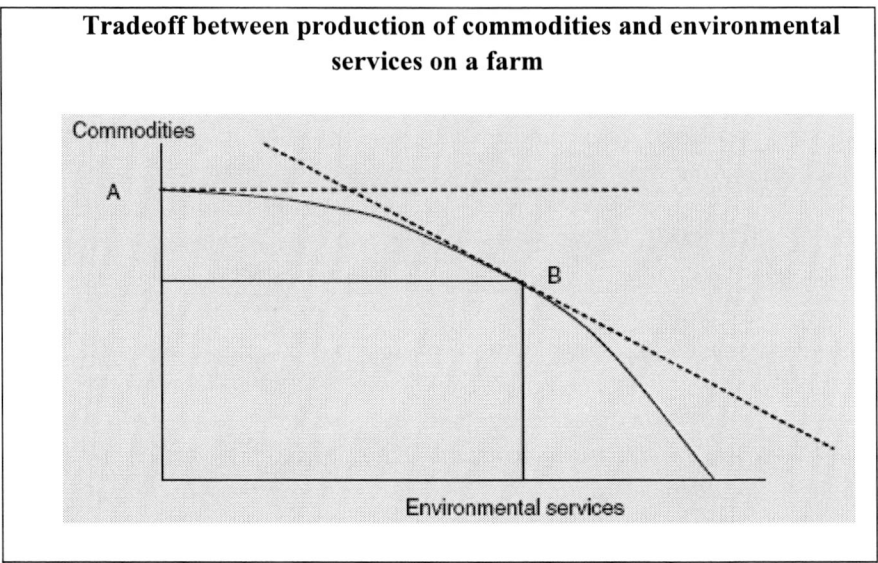

The point is that prices tell market participants how valuable one good or input is relative to another, making prices the most important piece of information driving decisions about production and consumption. If prices for environmental services under-represent their true value, fewer resources will be directed toward the production of environmental services than is socially optimal. The public-good nature of most environmental services is the primary reason that markets for them do not develop. Consequently, the price for most environmental services is zero.

Transaction Costs

Transaction costs are the costs of doing business. Parties must find one another and exchange information. They may also have to inspect or measure the good, draw up contracts, and consult with lawyers or regulators (Stavins, 1995). These actions require inputs of time or resources, costs that reduce the overall benefits expected from the transactions. If transaction costs are high relative to the value of the good, then exchange may have no benefits and a market could fail to develop.

Transaction costs associated with potential markets for environmental services are likely to be high. One issue with environmental services from agriculture is that they are often secondary to a farmer's primary activity of producing agricultural commodities; they are produced as externalities of agricultural production. It may be too costly for a farmer to learn about potential demand for an environmental service, develop a business plan, keep the necessary records, and integrate the new business into the traditional farming operations.

Environmental services, such as water quality and carbon sequestration, are diffi cult to measure. The monitoring necessary to measure these services is often expensive and may require intrusive visits to the farm.

Traditional farm commodities already have established systems for collection and distribution. Farm commodities are generally homogeneous, prices are established in centralized markets, and agricultural producers do not have to negotiate with each potential final buyer. On the other hand, environmental services tend to be unique for each farm, with no standard form of transaction. A farmer wishing to sell an environmental service may have to negotiate with each potential buyer, a potentially costly process. The same would be true for a buyer of environmental services, who may have to negotiate with many farmers.

Uncertainty

The performance of conservation practices in the production of environmental services is one of the most important sources of uncertainty in environmental markets. Uncertainty about the quantity and quality of services a farmer can produce affects both the demand and supply side of markets. Markets function best when information on the commodity is complete and readily available to all potential market participants. Environmental services, however, are often diffi cult to observe, such as the nutrient-fi ltering capacity of wetlands or the sequestration of greenhouse gases from adopting conservation tillage. Determining the quantity of services a farm can produce is, therefore, often left to estimation, based on farming practices and location. When information is missing, or otherwise inaccessible, potential customers may be reluctant to enter the market, or they may trade less. Uncertainty can also affect producers. Agricultural producers are reluctant to adopt management practices if potential returns are uncertain. Not knowing the quantity of environmental services that can be produced and sold would be a major impediment to entering a market for environmental services. Determining the amount and nature of the services a farm can produce can be costly, especially given their complex nature.

Institutional Barriers

Institutional barriers may prevent agricultural producers from selling an environmental service in existing markets. Agricultural producers may be unable to sell environmental services either by rule or because the rules that govern participation limit the supply of services a farm can provide in a market.

For example, participants in the Wetlands Reserve Program are prohibited from selling some environmental services created by wetland restoration paid for by taxpayers, including carbon sequestration, open space, and wetland services (for the purposes of mitigation) (USDA, Natural Resources Conservation Service, 2007a).

Some markets do not allow environmental services from agricultural sources because of a high level of uncertainty about the amount actually produced or about their long-term supply. For example, some markets for greenhouse gas reduction do not allow credits from sequestration in agricultural soils because of the risk of future carbon emissions due to changes in management (known as the permanence issue) (Ecosystem Marketplace, 2007a). One could argue that uncertainty would reduce the demand for such credits in a market anyway and be reflected in price, but some markets have chosen to take away the choice entirely.

Some water quality trading programs require agricultural producers who wish to sell credits to be practicing a minimum level of stewardship. Requiring a minimum level of stewardship to participate in the trading program prevents the lowest cost credits from being marketed, raising the overall price of credits for point sources. The requirement is also a barrier for some producers, discouraging them from entering the market. A producer may be unwilling to bear the cost of achieving the minimum level of stewardship before being eligible to sell credits.

While not necessarily a barrier, government programs can sometimes compete for producers' investment in environmental stewardship. Government conservation programs and markets for environmental services sometimes have common objectives and outcomes. For example, conservation programs and trading programs may compete with each other for pollution reductions from agriculture. If a farmer enrolls in a conservation program to reduce nitrogen runoff, the marginal cost of making additional environmental gains (beyond those funded by the conservation program) is higher. If the farmer then wishes to participate in a trading program, the cost of abatement credits is higher than it would have been otherwise. Agricultural producers with a history of heavy involvement in conservation programs may have a more difficult time competing in a market than if they had not been as involved. While the environmental service is still being provided, market forces are not guiding the allocation of resources.

SUMMARY

The public-goods nature of environmental services is the most important reason that markets for environmental services have not developed on their own. Transaction costs, uncertainty, and institutional barriers are also factors inhibiting markets. Government can use a variety of policy tools, including market mechanisms, to create incentives for farms to provide environmental services. The following chapter presents some examples of how market mechanisms have been used to spur the provision of environmental services, as well as steps government can take to promote the creation of sustainable markets.

Chapter 4

WHAT CAN WE LEARN FROM CURRENT MARKETS?

The previous chapter showed how characteristics of environmental services from agriculture prevent well-functioning markets from developing and hinder market function. As a result, the prices that convey information about the relative values of goods and services in well-functioning markets either do not exist in markets for environmental services or convey fl awed information. Can anything be done to "fi x" the system so appropriate information is conveyed to landowners who provide environmental services?

To obtain a clearer understanding of how markets can be used to help provide environmental services, this chapter takes a close look at five different markets. For each market, we describe the "good" that is being bought and sold, impediments to demand, impediments to supply, and steps taken by government and/or market participants to overcome those impediments. Since we are also interested in the extent to which markets for environmental services might become a signifi cant source of financial resources for stewardship on farms, we also explore the potential size of these markets. The five markets examined are water quality trading, carbon trading, wetland mitigation, wildlife, and eco-labels.

WATER QUALITY MARKETS

Agriculture signifi cantly affects water quality (chapter 2). Farmers and ranchers, for the most part, have little incentive to improve water quality. The primary U.S. water quality law, the Clean Water Act (CWA), regulates pollution only from point sources (for example, factories, sewage treatment plants, and large confined animal feeding operations). Voluntary approaches for controlling pollution from agriculture are the mainstay of Federal and State water quality improvement efforts. But benefits from water quality improvements occur mostly off the farm, and since they are public goods, few producers would voluntarily incur the costs of adopting management practices that improve water quality. How can a market for water quality be created?

One approach is emissions trading. Emissions trading is organized around the creation of discharge allowances, which is a time-limited permission to discharge a fixed quantity of pollutant into the environment. A discharge allowance has characteristics of a private good; it is rival and exclusive. Property rights are enforced by the regulatory agency managing the program.

A discharger (assumed to be a profit-maximizing firm) must own allowances to legally release pollutants. A regulatory agency creates demand for discharge allowances (and reduces pollution in regulated waterways) by restricting the number of allowances in a market. The regulatory agency first determines the maximum amount of discharge of a particular pollutant a watershed can absorb and still meet environmental quality goals. This becomes the emissions cap for the watershed. The cap is used to set discharge limits for each regulated firm operating within the watershed. Discharge allowances equal to the emissions cap are allocated to all regulated dischargers through an auction or some other means. By enabling allowances to be traded, a market is created that allocates discharges among regulated firms.

If a firm discharges more pollution than its holding of allowances during the year, it would be subject to fines and penalties. If a firm does not have enough discharge allowances, it can either reduce discharges or purchase allowances from other firms. If a firm discharges less than its holding of allowances, it can sell the excess. A firm will purchase allowances in the market if the price is less than its cost of reducing a unit of discharge. If a firm can reduce discharges at a cost lower than the price of an allowance, it will reduce emissions below its permit requirements and sell the excess allowances and earn a profit. If the market operates smoothly, it can achieve environmental goals at a lower cost than command and control regulations alone (Tietenberg, 2006). Firms with low pollution control costs will provide proportionately more pollution control, reducing total pollution control costs. A market allows maximum flexibility for firms in that a firm can meet its obligations by installing pollution control technology, adopting more effi - cient production technology, rearranging production processes, or purchasing credits (Ribaudo, Horan, and Smith, 1999). Emissions trading has been very successful in reducing the cost of regulations on sulfur dioxide emissions to the atmosphere from power plants (see box, "Trading

Can Reduce the Cost of Lowering Emissions"). This program is estimated to have exceeded environmental goals at a savings of over $1 billion compared with a regulatory approach that does not allow trading (Stavins, 2005).

In the textbook example of emissions trading, all market participants are regulated under the cap. In water quality trading programs, EPA allows regulated point sources to purchase credits from unregulated nonpoint sources, such as agriculture. Sources of credit outside the cap are known as offsets.

Water quality trading markets must meet some basic conditions in order for demand for credits from nonpoint sources to develop. Units of trade must be clearly defined, defensible ecologically and economically, consistently measured, and enforced by the regulatory agency (Boyd and Banzhaf, 2006). The commodity to be traded must be a single pollutant in a common form that is understood by market participants. The discharge point of purchase and sale must be environmentally equivalent to ensure that expected water quality gains are achieved. The timeframes for buyers and sellers of credits must be aligned, in that purchased reductions in discharge must be produced during the same period that a buyer was required to produce them. The supply of nonpoint credits must be in balance with the point sources' demand for credits, in that there are enough potential nonpoint credits to satisfy the needs of potential purchasers. Otherwise, trading with nonpoint sources would not be able to generate pollution control savings.

Experience with water quality trading programs highlights the problems with nonpoint-source-created credits and some of the steps that can be taken to address those problems. Since 1990, 40 water quality trading programs have been started in the United States; 15 include production agriculture as a potential source of credits for regulated point sources (table 4.1) (Breetz et al., 2004). To date, trades between point and agricultural nonpoint sources have occurred in only four programs: Piasa Creek (Illinois), Red Cedar River (Wisconsin), Southern Minnesota Beet Sugar, and Rahr Malting (both Minnesota). These trades appear to be cost effective. For example, in the trading program established for Rahr Malting, four nonpoint-source projects controlled phosphorus runoff at a cost of about $2.10 per pound (Breetz et al., 2004). Rahr Malting would have had to pay an estimated $4-$18 per pound of phosphorus reduced if it had installed pollution control equipment. However, supply-side and demand-side impediments seem to be preventing trades in most trading programs. Simply creating a private good related to water quality by itself is insufficient for generating market activity.

TRADING CAN REDUCE THE COST OF LOWERING EMISSIONS

Without trading, the regulated firm reduces discharges by 500 pounds at a cost of $25,000 (500 pounds at $50 per pound), and the farm does nothing.

With trading, the firm reduces discharges by 400 pounds at a cost of $20,000 (400 pounds at $50 per pound). The farm is willing to reduce discharges for a price of $15 per pound. The firm purchases 100 pounds of reduction from the farm at a cost of $1,500 (100 pounds at $15 per pound). The firm's costs have been reduced to $21,500 (a savings of $3,500). The farm reduces discharges by 100 pounds at an actual cost of $1,000 (100 pounds at $10 per pound). The farmer receives a payment of $1,500 from the firm, so he or she actually realizes a profi t of $500 for trading with the firm.

The total cost of reducing pollution (not considering profi t to the farmer) has been reduced from $25,000 to $21,000.

Example: Firm discharge limit, no trading

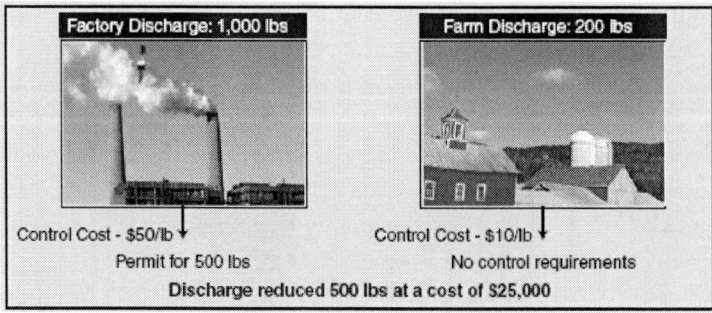

Example: Firm discharge limit, with trading

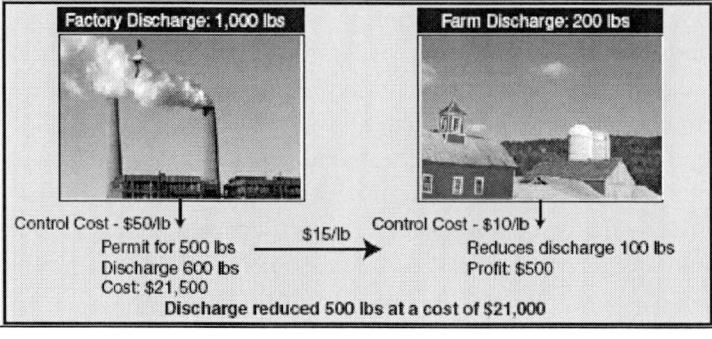

Table 4.1. Water quality trading programs that include agriculture

Project	Pollutant traded	Trades
		Number
Cherry Creek, CO	Phosphorus	0
Lower Boise River, ID	Phosphorus	0
Piasa Creek, IL	Sediment	1
Acton, MA	Phosphorus	0
Massachusetts Estuaries Project	Nitrogen	0
Kalamazoo River, MI	Phosphorus	0
Rahr Malting, MN	Phosphorus	4
Southern Minnesota Beet Sugar, MN	Phosphorus	579
Tar-Pamlico, NC	Nitrogen, phosphorus	0
Clermont County, OH	Nitrogen, phosphorus	0
Great Miami River, OH	Nitrogen, phosphorus	0
Conestoga River, PA	Nitrogen, phosphorus	0
Fox-Wolf Basin, WI	Phosphorus	0
Red Cedar River, WI	Phosphorus	22
Chesapeake Bay Watershed	Nitrogen, phosphorus	0

Source: Breetz et al., 2004.

Issues in Demand for Credits from Agriculture

The source of demand for credits in any trading program is a regulation that establishes a cap on discharges that is below current levels. In the case of water quality, the Total Maximum Daily Load (TMDL) provision of the Clean Water Act is the legal mechanism that establishes a cap on pollution discharges in impaired watersheds. Without an effective or binding cap, regulated sources have no reason to seek credits in a market. Ineffective caps on point-source (regulated) dischargers are cited as the reason for lack of demand for nonpoint-source credits in three trading programs and may be a problem in others (Breetz et al., 2004).

One of the requirements of trading is the equivalency of credits; ideally, point-source purchases of credits in a market have the same impact on water quality as if the firm reduced discharges itself. This equivalency ensures that water quality goals are actually met. Establishing equivalency between point and nonpoint sources must account for two factors—agricultural practice effectiveness and location relative to the point source.

The effectiveness of a best management practice (BMP) depends on site-specifi c conditions, implementation, and how well it is maintained (Mid-Atlantic Regional Water Program, 2006). Uncertainty about such performance is a major stumbling block with point-nonpoint trading. If a regulated point source is legally responsible for achieving a particular discharge goal, the uncertainty about credits generated by nonpoint sources may make them an unattractive option. A point source's control strategy is generally a long-term decision, and it may be unwilling to rely on an uncertain source of credits because of the decision's inherent irreversibility (McCann, 1996). These factors may push point sources toward providing their own internal emission controls or trading with other point sources, rather than relying on nonpoint credits. Measurement problems were cited as obstacles in several existing trading programs (Breetz et al, 2004). Uncertainty about practice performance can be addressed in three ways. One is to conduct research on the performance of practices under different conditions. USDA's Agricultural Research Service (ARS) conducts extensive research on the environmental performance of production practices and could provide information that reduces uncertainty in trading programs. Some water quality trading programs use simulation models to predict the performance of practices. Research and model development are costs that are generally borne by the public.

A second approach is to address the liability issue. A number of trading programs have created a reserve pool of credits that can be used by regulated point sources when an offset project fails to produce the expected number of credits. This pool could increase the willingness of point sources to trade with nonpoint sources. Rules that grant some leeway for point sources that purchase nonpoint-source offsets would also encourage pointnonpoint trading.

A third approach for addressing practice uncertainty is an uncertainty ratio. An uncertainty ratio is a type of trading ratio that generally requires more than one unit of nonpoint-source discharge reduction to offset one unit of point-source discharge. Uncertainty ratios in water quality trading programs generally range from 2:1 to 5:1, which means that a point source would have to purchase up to five units of pollutant reduction from a nonpoint source in order to ensure that its single unit of discharge is "covered" (Conservation Tillage Information Center, 2006). While providing assurance that the nonpoint-source reduction provides the expected gain in water quality, a trading ratio increases the effective price of nonpoint credits, thereby reducing point sources' demand for them. Research on practice performance could reduce this ratio, making nonpoint-source credits less costly to point sources.

Establishing equivalency between nonpoint offsets and point-source discharges also must take into account the location of nonpoint sources relative to the point source. Since equivalency is measured at the point source, the fate of pollutants when they leave a field must be considered as they move downstream. Take two fields, one close to the point source and the other much farther upstream. Identical reductions in nitrogen runoff at the two fields would affect water quality differently, as measured at the point source, due to biophysical activity along the way: the closer the source, the greater the effect. This difference must be accounted for when potential trades are constructed. A delivery or location ratio is another type of trading ratio, accounting for the location in the watershed of the nonpoint source relative to the point source: the smaller the distance, the smaller the ratio. While providing assurance that the nonpoint-source reduction provides the expected gain in water quality, a delivery ratio increases the effective price of nonpoint credits from farms located farther from the point source, thereby reducing point sources' demand for them.

Another issue facing point sources' demand for nonpoint credits is the cost of finding trading partners. Because farms are generally widely distributed across a watershed and each may be capable of producing a relatively small number of discharge credits, the transaction costs for point sources of identifying enough willing trading partners to satisfy their permits may discourage them from seeking trades. Some markets have developed formal clearinghouses that assemble information from both buyers and sellers, making it easier for potential trading partners to find each other (Breetz et al., 2004). Third-party aggregators are also used in several markets to assemble credits from nonpoint sources. Aggregators then market the credits to potential purchasers. Both government and nongovernment organizations are playing roles of clearinghouse and aggregator.

Issues in Supply of Credits from Agriculture

Some of the impediments to the formation of trading markets fall on the supply side. Farm runoff is not regulated under the Clean Water Act, so producers are not compelled to actively seek trading partners. The expected returns from trading may not adequately compensate for the type of inspection and scrutiny the farm may receive if it enters into a trading program. Evidence from existing programs suggests that producers may also avoid trading programs because of a fear that entering into a trade is an admission that their farms pollute, exposing them to citizen complaint or future regulation (King and Kuch, 2003; King, 2005; Breetz et al., 2004).

Farmers may be uncertain about the number of credits they can reasonably expect to produce, making it difficult for a producer to determine whether it is financially beneficial to enter a market. Models and other tools could help farmers reduce this uncertainty. An example of this type of information source is the World Resources Institute's NutrientNet (World Resources Institute, 2007). This online tool can function as an information source for farmers. Configured to a specific watershed, NutrientNet allows registered users to evaluate different trading options and assesses the combination of practices that works best for a farm with a particular set of resource characteristics.

Another tool currently under development is the NRCS/EPA Nitrogen Trading Tool (NTT). NRCS developed the NTT, in cooperation with ARS and EPA, as an online tool to help farmers determine how many potential nitrogen credits they can generate on their farms and sell in a water quality trading program (Gross et al., 2008). It allows a farmer to enter geographic, agronomic, and land use information to estimate baseline nitrogen loadings and changes in management practices or land use to calculate nitrogen load reductions that are the basis for credits in a trading market. Tools such as NutrientNet and the NTT can also reduce uncertainty on the demand side, if the model results are found to be reliable estimates.

A trading program may specify a set of practices eligible for producing credits to those for which performance data are readily available (Conservation Tillage Information Center, 2006). While simplifying the programs' problem of evaluating potential trades, it limits the choices a farmer may make in supplying credits. If the list of practices does not appeal to a farmer, he or she may decide not to participate.

Another supply-side issue arises when producers also participate in conservation programs, such as USDA's Environmental Quality Incentives Program (EQIP). Most trading programs do not allow producers receiving financial assistance for water-quality-protecting management practices through Federal programs to sell the subsequent water quality improvements as credits to point sources. An additional payment from a point source would not improve water quality beyond what the Government has already paid for. If farmers pay part of the cost of the practice out of their own pockets, one solution might be to allow a portion of the credits to be sold.

Some trading programs require a minimum level of stewardship before credits can be generated. For farms without "acceptable" management practices, credits cannot be created until the base level of environmental performance is attained. This requirement prevents the lowest cost credits from farms that have not adopted acceptable practices from being sold on the market, unless the returns from selling credits is so high that both the initial investment to achieve the baseline and the subsequent management costs can be covered. The bottom line is that the supply of low-cost credits is reduced, which has the effect of increasing the price regulated firms must pay.

Coordination of conservation programs with trading programs is one solution. USDA conservation programs, such as EQIP, could be targeted to producers who are not meeting the minimum level of stewardship to encourage them to participate in a trading program. The number of producers likely to participate in the trading program would increase, raising the potential supply of credits. However, average costs of credits would still be higher than if a stewardship-based baseline had not been used.

Producers may also face high transaction costs when trying to find trading partners. A farmer has to consider the type, amount, and timing of pollutant reductions generated on the farm and determine if they match the type, amount, and timing of pollutant reductions needed by regulated dischargers (Conservation Tillage Information Center, 2006). Unfamiliarity with the regulated community and the negotiation process could discourage producers from participating in a trading program. Third-party aggregators can play a role in addressing this issue and are being used in several projects. Trading programs have also established outreach programs to educate farmers about the opportunities that trading might offer and how to participate.

Future Role for Agriculture in Trading Programs?

USDA's interest in water quality trading (and other markets for environ-mental services) is based largely on the potential level of financial resources from private sources for conservation on farms. A question we examine is the extent to which water quality trading could provide enough financial assistance to producers to address a signifi cant amount of agricultural nonpointsource pollution in impaired watersheds, assuming that demand and supply impediments could be overcome. We use a simple screening procedure to identify watersheds where demand for water quality credits by point sources may be high and where agriculture can provide enough credits to meet that demand, assuming that nonpoint sources of pollution, such as agricultural producers, remain unregulated.

Data and Analysis

Our goal is to identify watersheds that could support an active trading market with agriculture as a supplier of credits. To do so, we first identifi edwatersheds where nutrient loadings are identified as a problem. Our analysis includes the 2,111 eight-digit Hydrologic Unit Code (HUC) watersheds of the contiguous United States. Data on nutrient impairment were obtained from EPA's 303(d) list of State-reported impaired waters (U.S. EPA, 2007a). With these data, we identified 710 HUCs containing water bodies impaired by nutrients—i.e., either nitrogen (N) or phosphorus (P).

We then identified watersheds where agriculture is likely to be a credit supplier. Because point sources may be required to purchase three or more credits from nonpoint sources for each unit of discharge, we assume that only watersheds where agriculture contributes a large portion of total nutrients— greater than 50 percent —might develop an active credit market where signifi - cant revenue for water-quality-enhancing practices flows to the agricultural sector. Finally, to ensure sufficient demand for nonpoint-source credits, we consider only watersheds where point sources contribute at least 10 percent of loadings. Estimates of nutrient loadings from point sources, agricultural nonpoint sources, and other nonpoint sources in each HUC were obtained from the U.S. Geological Survey (2000).

An important aspect of the potential price of credits from agriculture is the level of nutrient management that is part of the baseline (from which created credits are calculated). As discussed earlier, the cost of supplying a water quality credit is likely to be lower in watersheds with a lower percentage of cropland under a nutrient management plan (NMP). Data on the amount of cropland already covered by a NMP implemented with assistance from USDA in each HUC during 2004-06 were obtained from the NRCS Performance Results System (USDA, NRCS, 2007c).

Results

Agriculture is the primary source of nutrient loadings in most of the 710 impaired HUCs. Agriculture is responsible for 91-99 percent of N loadings in 68 percent of the impaired HUCs (figure 4.1). Similarly, agriculture is responsible for 9 1-99 percent of P loadings in 52 percent of HUCs (fi g. 4.2). We expect relatively low demand for agricultural credits (as a share of total agricultural discharges) by point sources in these watersheds because of the predominance of nonpoint-source loadings. Even though point sources may benefit from a plentiful supply of credits, only a small percentage of agriculture's contribution to pollution will be addressed through management practices funded by point sources.

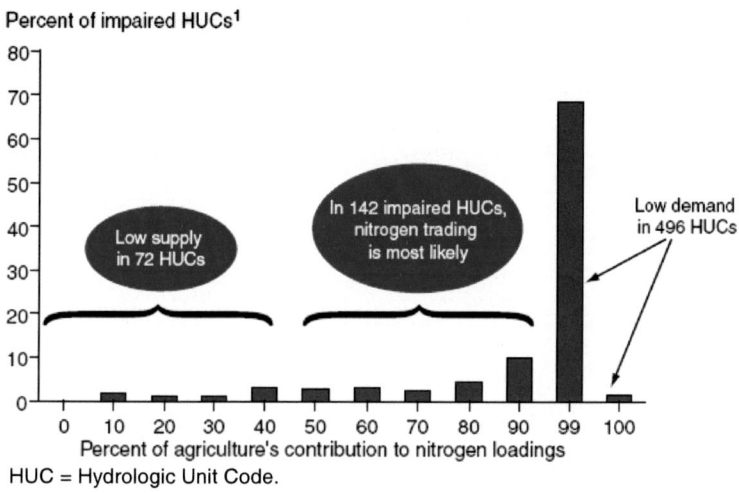

^{1}Number of impaired HUCs = 710.
Sources: USDA, ERS analysis of Environmental Protection Agency and U.S. Geological Survey data.

Figure 4.1. Agriculture's contribution to within-HUC nitrogen loadings.

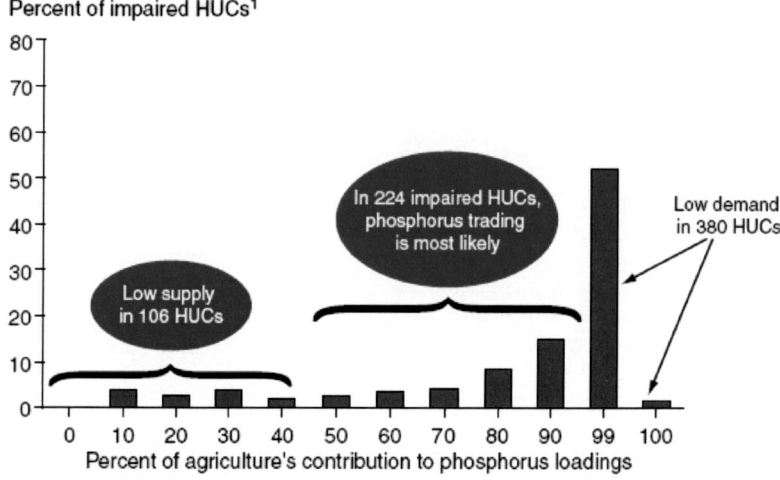

HUC = Hydrologic Unit Code.
^{1}Number of impaired HUCs = 710.
Sources: USDA, ERS analysis of Environmental Protection Agency and U.S. Geological Survey data.

Figure 4.2. Agriculture's contribution to within-HUC phosphorus loadings.

Agricultural contributions of N and P ranging between 50 and 90 percent are found in 142 and 224 of the impaired HUCs, respectively. We believe that demand and supply of credits is more balanced in these watersheds, which is necessary for an active market. Figures 4.3 and 4.4 show the spatial distribution of HUCs that meet our screening criteria for nitrogen and phosphorus. There are about 322,000 farms (15 percent of all U.S. farms) in the watersheds where phosphorus trading markets may be viable, and about 175,000 farms (8 percent) in the watersheds where nitrogen trading markets may be viable (table 4.2).

In terms of the cost of credits that agriculture might supply, no HUC had more than 22 percent of its cropland under a NMP, and most had less than 5 percent, which suggests that the level of NMP adoption would influence the level of trading in very few HUCs.

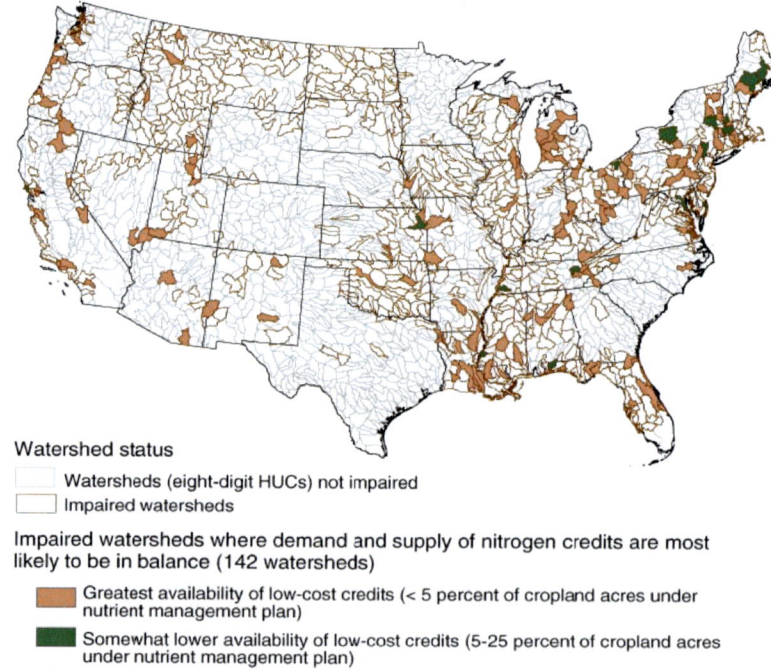

Watershed status

  Watersheds (eight-digit HUCs) not impaired
  Impaired watersheds

Impaired watersheds where demand and supply of nitrogen credits are most likely to be in balance (142 watersheds)

  Greatest availability of low-cost credits (< 5 percent of cropland acres under nutrient management plan)
  Somewhat lower availability of low-cost credits (5-25 percent of cropland acres under nutrient management plan)

HUC=Hydrologic Unit Code.

Note: None of the 142 watersheds have over 25 percent of cropland acres under nutrient management plan.

Source: USDA, ERS analysis of Environmental Protection Agency, U.S. Geological Survey, and Natural Resources Conservation Service data.

Figure 4.3. Nitrogen credit trading opportunities: Impaired watersheds where active trading is most likely.

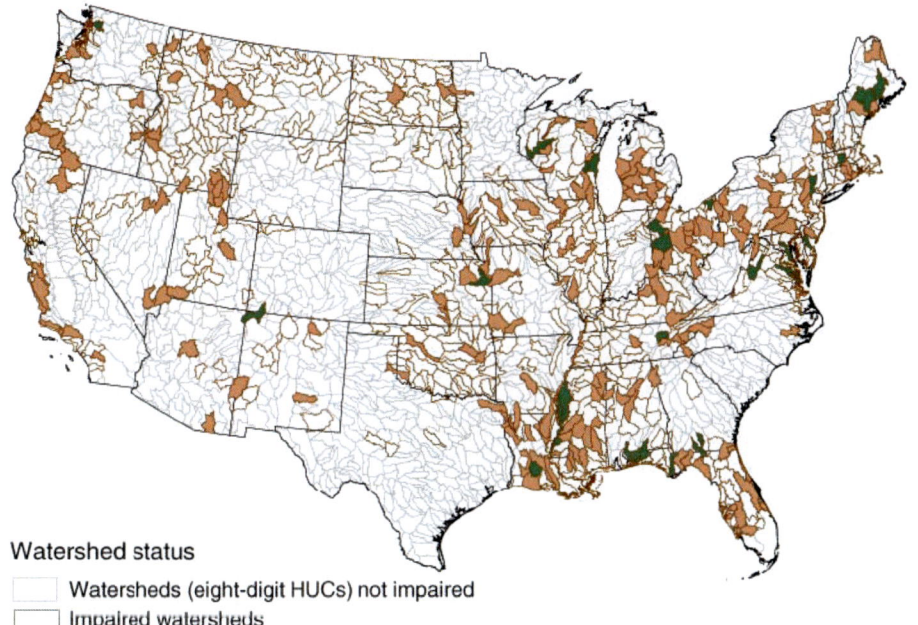

Watershed status
☐ Watersheds (eight-digit HUCs) not impaired
☐ Impaired watersheds

Impaired watersheds where demand and supply of phosphorus credits are most likely to be in balance (224 watersheds)

■ Greatest availability of low-cost credits (< 5 percent of cropland acres under nutrient management plan)

■ Somewhat lower availability of low-cost credits (5-25 percent of cropland acres under nutrient management plan)

HUC=Hydrologic Unit Code.
Note: None of the 224 watersheds have over 25 percent of cropland acres under nutrient management plan.
Source: USDA, ERS analysis of Environmental Protection Agency, U.S. Geological Survey, and Natural Resources Conservation Service data.

Figure 4.4. Phosphorus credit trading opportunities: Impaired watersheds where active trading is most likely.

Trading could occur in any HUC where point sources are required to reduce nutrient loadings and are allowed to offset their discharges with reduction from farms. Most of trades that have actually occurred are single point sources that offset pollution through contracts with multiple producers. However, these results indicate that trading is not likely to be a major source of conservation assistance, even if impediments to trading are overcome. Relatively few impaired watersheds are in "balance," in that the potential demand for nonpoint-source credits is high enough to spur an "active" market with nonpoint sources. Although trading may represent an important source of conservation funding in some local areas, USDA will likely remain the primary source of financial assistance for water quality protection on farms, assuming that nonpoint sources remain unregulated.

Table 4.2. Farms and income in watersheds where trading most likely

Indicator	Nitrogen		Phosphorus	
	<5% NMP	5-25% NMP	<5% NMP	5-25% NMP
Farms (number)	156,846	174,724	281,191	321,654
Crop sales ($1,000)	7,085,235	7,577,431	12,733,629	14,246,215
Livestock sales ($1,000)	4,997,249	5,718,705	10,455,075	12,411,055
Net cash income ($1,000)	2,632,522	2,864,816	5,315,743	6,048,930

NMP = Nutrient management plan.
Note: The United States has about 2.1 million farms.
Source: 2002 Census of Agriculture.

GREENHOUSE GASES AND AGRICULTURE

Concerns about global climate change have led to various strategies for reducing greenhouse gases (GHG) in the atmosphere. Most strategies combine reductions in emissions of GHG with sequestration (long-term removal of GHG from the atmosphere). One policy approach is to create markets for greenhouse gas reductions. As described in chapter 2, agriculture is both a source and sink for greenhouse gases, and producers might benefi t in such markets by reducing greenhouse gas emissions, or by sequestering carbon in the soil or in biomass.

One factor in favor of developing an active market is the worldwide potential of such a market. Reducing greenhouse gas emissions or sequestering carbon have the same benefi t no matter where they occur geographically, which means GHG reduction credits have many potential buyers and sellers, a necessary condition for an active market.

Issues in Demand for GHG Reductions

There are two primary scenarios for "creating" demand for GHG reductions (Council for Agricultural Science and Technology, 2004):

1. Regulatory cap and trade markets that set emission limits.
2. Voluntary markets driven by the following:
   - Consumer willingness to pay to reduce their carbon "footprint."
   - Firms wishing to show themselves as responsible environmental actors.
   - Firms wishing to gain control of low-cost alternatives that may be used to comply with future emission limitations (speculation) (Butt and McCarl, 2004).

Regulatory Markets

Regulatory markets create a property right for GHG reductions, in the form of tradable credits, much like the discharge allowance in water quality markets. The European Union Emissions Trading Scheme (EU ETS) is the world's largest market in greenhouse gas emissions. It was established primarily to help the 25 EU member states achieve their Kyoto Protocol targets (Ecosystem Marketplace, 2007b). The program established a mandatory cap and trade program for carbon dioxide in 2005 that did not include carbon sinks.

Several State and regional cap and trade programs have recently been approved in the United States to reduce GHG emissions. The Oregon CO_2 Standard, established in 1997, is the only State-level program currently underway. It requires new power plants to reduce emissions to 17 percent below those of the most effi cient plant (Ecosystem Marketplace, 2007c). Affected companies have the option of meeting this requirement by financing carbon offset projects through the Climate Trust, a nongovernmental organization established to seek and finance offset projects and to verify offset credits. While the Oregon program rules do not place any limitations on the geographic location or types of CO_2 offset projects, Climate Trust does not accept offsets from sequestration in agricultural soils (Climate Trust, 2007).

Member States of the Regional Greenhouse Gas Initiative (RGGI) (consisting of 10 Northeastern and Mid-Atlantic States), Washington, and California are developing cap and trade programs for reducing GHG emissions, primarily from power plants. Rules are still being developed for these programs, but Climate Trust is already seeking carbon offset projects to meet future demand from RGGI (Climate Trust, 2007).

Voluntary Markets

Voluntary markets are currently the greatest source of demand for GHG reduction credits in the United States. The Chicago Climate Exchange (CCX) is a voluntary cap and trade program covering emission sources from the United States, Canada, and Mexico and offset projects from these countries and Brazil. While joining CCX is voluntary, members make a legally binding commitment to meet annual greenhouse gas emission reduction targets (Ecosystem Marketplace, 2007a). Members agree to annual reductions that will reduce their emissions of greenhouse gases by 6 percent below the average of their 1998-2001 emissions baseline by 2010. Each member can meet its commitment through internal reductions, by purchasing allowances from other members, or by purchasing credits from emissions reduction projects. CCX issues tradable Carbon Financial Instrument contracts to owners or aggregators of eligible projects on the basis of sequestration, destruction, or displacement of GHG emissions. Eligible projects include agricultural methane, landfill methane, coal mine methane, agricultural and rangeland soil carbon, forestry, and renewable energy. As of July 2007, the price of a CO2 equivalent[2] (CO_2e) was $3.25 per ton (Ecosystem Marketplace, 2007d). In contrast, carbon offsets in the European Union's Emissions Trading Scheme were trading at $30.60 per ton CO_2e (Ecosystem Marketplace, 2007d). This difference reflects the fact that the CCX is voluntary, while the EU ETS is not, and agricultural soil sinks are not recognized as a source of permanent carbon reductions by EU ETS. Since its inception in 1997, the CCX has traded almost 24 million metric tons of CO_2e.

The CCX addresses the issue of the cost of finding potential offsets from a geographically dispersed sector through the use of third-party aggregators. Aggregators create, aggregate, register, and trade certified carbon credits to buyers in the CCX. Fifty-three offset aggregators are members of the CCX and include farm groups (e.g., Iowa Farm Bureau) as well as private corporations (Chicago Climate Exchange, 2007).

An important question is why a firm would voluntarily enter into a legally binding commitment to reduce its carbon emissions. Some of the benefits include the following (Chicago Climate Exchange, 2007):

- Capture gains and manage risks in the growing carbon market.

---

[2] A carbon equivalent is an internationally accepted measure that expresses the global warming potential of greenhouse gases in terms of the amount of carbon dioxide that would have the same global warming potential.

- Acquire cutting-edge measurement and trading skills that will be needed as markets develop.
- Demonstrate strategic vision on climate change to shareholders, rating agencies, customers, and citizens.
- Gain leadership recognition for taking early, credible, and binding action to address climate change.

A purely voluntary retail market for carbon offsets has developed for individuals, businesses, and other institutions that find the concept of being "carbon neutral" an attractive one. Approximately 35 retail offset providers currently offer "carbon neutrality" for a fee to consumers and businesses. These retailers fund projects that are intended to offset GHG emissions from cars, airplanes, and special events, such as concerts and weddings. Offset projects include a variety of actions, including methane capture from animal feeding operations and landfills, reforestation, developing renewable energy, and improving energy efficiency. Some retailers purchase reductions on the CCX rather than directly funding projects. Retailers are currently charging from $4 to $35 per ton of CO2e.

Demand in this retail market is largely unknown. Relatively little information is available regarding the volume of trades or the composition of volume by project type (Trexler, Koslof, and Silon, 2006). In addition, little research has been conducted on consumers' willingness to pay to be carbon neutral. The good being sold does not have private-good characteristics, so free riding is an issue for market development.

Another implication of the newness of these markets is that they currently have no accepted standards for what qualifies as an offset for making consumers carbon "neutral." Because the commodity is intangible, it is very difficult for consumers to differentiate between high-quality and low-quality offerings based on the information provided by retailers (Trexler Climate + Energy Services, 2006). In addition, there are no industry quality standards for offsets, no reliable certification process for retailers, and no effective disclosure and verification protocols. Consumers who are willing to pay may be reluctant to enter such markets because of this uncertainty. Such uncertainty reduces overall demand, keeps prices low, and stifles market growth.

A related issue affecting demand in all carbon markets is whether nonagricultural entities will purchase carbon credits from agriculture. The quantity and permanence of carbon sequestration on agricultural soils is less certain than for other types of GHG reductions (Zeuli and Skees, 2000). The amount of carbon sequestered in agricultural soils is determined by the interaction of soils, climate, land use, crop rotation, fertilizer management, and other management practices

(Council for Agricultural Science and Technology, 2004). An accurate measurement of sequestration rates has to be made locally rather than relying on regional estimates. Demand by nonagricultural interests in this regard is not clear (McCarl and Schneider, 2000; Zeuli and Skees, 2000). Emphasis in many offset projects in both regulatory and voluntary markets is on permanent, easy-to-measure offsets, such as methane capture and destruction.

Research can address uncertainty issues surrounding potential sequestration of soil carbon. GRACEnet is an ARS project for estimating net GHG emissions of current agricultural systems and the impacts of alternative management (USDA, ARS, 2007). It will reduce uncertainty about how agricultural management might alter the amount of GHG emitted to the atmosphere by identifying the best regionally specifi c management practices for increasing soil carbon and reducing the net global warming potential of greenhouse gases emitted by agriculture. It will also provide a scientifi c basis for possible carbon credit and trading programs.

Issues in Supply of Carbon Sequestration

Carbon sequestration by agriculture is enhanced under management systems that (1) minimize soil disturbance and erosion, (2) maximize the amount of crop-residue return, and (3) maximize water and nutrient use effi ciency in crop production (Council for Agricultural Science and Technology, 2004). Changes in cropland management include adopting conservation tillage and residue management, improving crop rotations and cover crops, eliminating summer fallow, improving nutrient management, using organic manure and byproducts, and improving irrigation management (Lewandrowski et al., 2004). Land use changes include converting cropland to forests, perennial grasses, conservation buffers, and wetlands.

The potential supply is dictated partly by standards set by individual markets. A trading program's success depends on the agricultural sinks' ability to offer management practices that are visible and have a predicted effectiveness within acceptable degrees of certainty (McCarl and Schneider, 2000). The Chicago Climate Exchange limits cropland eligibility for carbon credits from conservation tillage to soils that have been evaluated for that purpose (Chicago Climate Exchange, 2007). The CCX also limits rangeland eligibility to regions where research on soil sequestration is available. Research programs, such as the one associated with GRACEnet, could reduce uncertainty about the potential for soils and management to sequester carbon, increasing the potential supply of offsets.

Such research also makes it easier for farmers to estimate expected returns from entering a market for carbon offsets.

A program that can help reduce uncertainty and transactions costs in greenhouse gas mitigation markets is the U.S. Department of Energy's revised Voluntary Greenhouse Gas Reporting Registry. The revised Registry, also known as the 1605b program, is a voluntary program for reporting GHG emissions to the Federal Government.[3] Participants can establish a record of emissions and emissions reductions that will be deemed "credible" over the widest possible range of potential uses (U.S. Department of Energy, 2007). The Registry provides guidance, tools, and standardized methodologies for estimating greenhouse gas emissions and removals. Possible benefi ts of the Registry include enhancing participants' ability to take advantage of future Federal climate policies in which emission reductions have value and helping agriculture and forest entities take advantage of State- and private-sectorgenerated opportunities to trade emission reductions and sequestered carbon.

We use a 2004 study by Lewandrowski et al., to get an idea of how farmers might respond to a price of $3.35 per ton of CO2e (price as of June 14, 2007, on the CCX) that is paid for gross reduction sequestration (not accounting for potential increases in GHG emissions elsewhere on the farm). Based on the study results, farmers in the 48 coterminous States would shift about 2.3 million acres of cropland to forest, about 11.5 million acres of grazing land to forests, and about 80 million acres of conventional tillage to conservation tillage. Net farm income for all farms would increase about 0.9 percent. These results assume no transaction costs or uncertainty and adequate demand to purchase all the credits farmers can sell at that price.

That the estimated levels of land-use changes have not occurred is due to a limited number of purchasers, transaction costs, uncertainty, and higher commodity prices than are used in the analysis.

What happens if payments are based on net sequestration rather than on gross sequestration? Unlike many commodities, what matters is not the flow of product (such as consumable ears of corn) but the stock (such as the amount of CO2 in the atmosphere). GHG trades outside the United States often require that emission reductions on a project site not lead to increases elsewhere (Butt and McCarl,

---

[3] The Registry was first created by paragraph 1605(b) of the Energy Policy Act of 1992. In February 2002, President Bush directed the Secretaries of Energy, Agriculture, and Commerce and the Administrator of the Environmental Protection Agency to recommend improvements to the program. The revised program stresses comprehensive (all GHG sources and sinks) and continuous (yearly) reporting, transparency in estimating emissions, and use of standardized estimation methods.

2004). For example, if a farmer retires 1 acre of cropland for production simply to switch his or her production efforts to an idle acre, there is no net benefit on the "stock" of carbon in the atmosphere (known as leakage). In the analysis described above, price increases for farm commodities provide an incentive to bring more land into production. Some of the benefits from sequestration are lost because of increased emissions on the farm. For the same price of $3.35 per ton of CO2e, if farmers are debited for changes in land uses and production practices that increase carbon emissions, the acres shifting to forests are the same but far fewer acres shift to conservation tillage—only 7 million acres compared with 80 million if payments are based on gross sequestration (Lewandrowski et al., 2004). Payments based on net sequestration are less attractive to farmers.

Supply of GHG reduction credits could be affected by how carbon markets and conservation programs are coordinated. Land retirement programs, such as the Conservation Reserve Program and Wetlands Reserve Program, and working lands programs, such as the Environmental Quality Incentives Program and Conservation Security Program, provide financial incentives for management practices that could be eligible for producing credits on the CCX. Unlike water quality trading programs, which often disallow credits produced through conservation programs, the CCX does allow credits produced via projects subsidized by conservation payments. Producers have an extra incentive to enroll in USDA conservation programs if they can also sell credits to the CCX. However, future carbon trading programs may not allow this "double dipping" because of questions about additionality.[4]

Farmer participation in carbon markets is also influenced by their willingness to accept market requirements. A shift to reduced- or no-till conservation practices could increase variations in net returns and discourage participation. The price of a carbon credit would have to be high enough to account for the increased uncertainty for a farmer to participate. Required practices may have management characteristics that do not mesh well with the rest of the farm, discouraging participation in the market (McCarl and Schneider, 2000). Producers may also be unwilling to make the long-term commitment that sequestration projects often require to be effective. This unwillingness may be exacerbated by tenure arrangements, which are generally for shorter periods and do not support long-term planning.

The way projects address uncertainty can affect the supply of credits. For example, each CCX project must place 20 percent of eligible carbon offsets in a

---

[4] [3]Additionality refers to emission reductions that are in addition to business-as-usual. They would not have occurred without the program.

reserve pool to provide coverage in the event the project fails to produce projected offsets (Chicago Climate Exchange, 2007). Offsets in the pool are held by the farmer and, if not needed, can be released for sale at the end of the accounting period. On the one hand, such a pool enhances trades by reducing uncertainty on the demand side. However, the uncertainty that this requirement imparts on expected income could discourage some producers from initiating a project.

Issues in the Supply of Methane Capture

Animal feeding operations are a potential source of methane emission reductions that are eligible in all carbon markets as offsets. Capturing methane and burning it, with or without the production of energy, reduces net GHG emissions. Eligible agricultural methane collection/combustion systems include covered anaerobic digesters, complete-mix, and plug-fl ow digesters. Methane emission reductions on animal feeding operations have been encouraged by assistance programs, such as EPA's AgSTAR, but the emergence of this market has improved the benefi ts to farmers of installing such systems.

Methane recovery systems are most effective for confi ned livestock facilities that handle manure as liquids or slurries, such as dairy and swine. EPA estimates that about 6,900 dairy and swine operations could benefi t financially by installing anaerobic digesters (U.S. EPA, AgSTAR, 2006). These include dairy operations with more than 500 head, swine operations with more than 2,000 head if using a system other than deep pit, and swine operations with more than 5,000 head if using deep-pit storage. Financial benefi ts include sale of carbon offsets to the various carbon markets and/or sale of electricity generated on the farm from captured methane. EPA estimates that it is technically possible for anaerobic digesters to reduce GHG emissions by 30 million metric tons CO2e per year (U.S. EPA, AgSTAR, 2006). For the sake of comparison, managed livestock waste emits about 51 million metric tons CO2e per year in the United States (USDA, Offi ce of the Chief Economist, 2007). About 90 operations have installed systems and are reducing GHG emissions by about 30,000 metric tons CO2e per year. How much of this is being sold on credit markets is not known. A number of factors affect whether animal operations would enter the GHG market. Digesters are very expensive, require a high level of management skill, and should be customized for each farm. Additionally, maintenance costs can be high.

WETLANDS' ENVIRONMENTAL SERVICES AND AGRICULTURE

Agriculture has traditionally had a profound effect on the supply of wetland services. Wetlands are rich ecosystems that provide a multitude of environmental services (see chapter 2). The number, mix, and quality of services in each bundle vary across wetlands, depending on their size and type, weather and climatic conditions, surrounding environment, and other factors.

As reported in chapter 2, wetland losses, primarily to agriculture, have been extensive (figure 4.5). Wetlands were drained without landowners considering their value; wetland services are public goods for which markets do not exit. Also, until the 1980s, USDA provided financial support for draining and filling wetlands, further tilting economic incentives toward reducing wetland services.

Demand for wetland services is expressed, indirectly, through purchases by government and nongovernment entities trying to preserve wetlands. Federal, State, and local governments act on the public's demand for wetland services by implementing programs and regulations that create and preserve wetlands and restrict wetland losses.

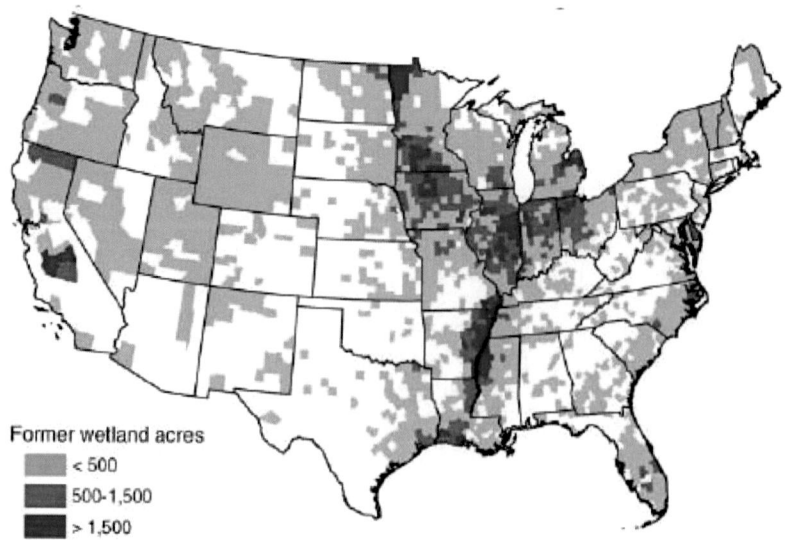

Source: USDA, ERS analysis of USDA's National Resources Inventory data (USDA, Natural Resources Conservation Service and Iowa State University, 200).

Figure 4.5. Country-level estimates of converted wetland acreage.

One program, USDA's Wetlands Reserve Program (WRP), has restored more wetlands than any other public effort. The WRP restores wetlands on agricultural lands and purchases easements on these lands. By the end of 2005, the WRP had enrolled over 1.9 million acres. WRP easements are found in every State (figure 4.6).

Nongovernment entities purchase wetland easements to preserve and increase the availability and quality of wetland environmental services. The magnitude of these purchases is diffi cult to gauge. Two of the more sizable organizations involved are The Nature Conservancy and Ducks Unlimited. The Nature Conservancy (TNC) uses a wide range of approaches to meet its mission, including the purchase of land and easements. Wetlands are included in over 4,000 projects encompassing more than 2.5 million acres and valued at nearly $2.6 billion.[5] In 2005, Ducks Unlimited controlled nearly 222,000 acres under easements or deed restrictions for the purpose of restoring and improving wetlands critical for waterfowl (Ducks Unlimited, 2007).

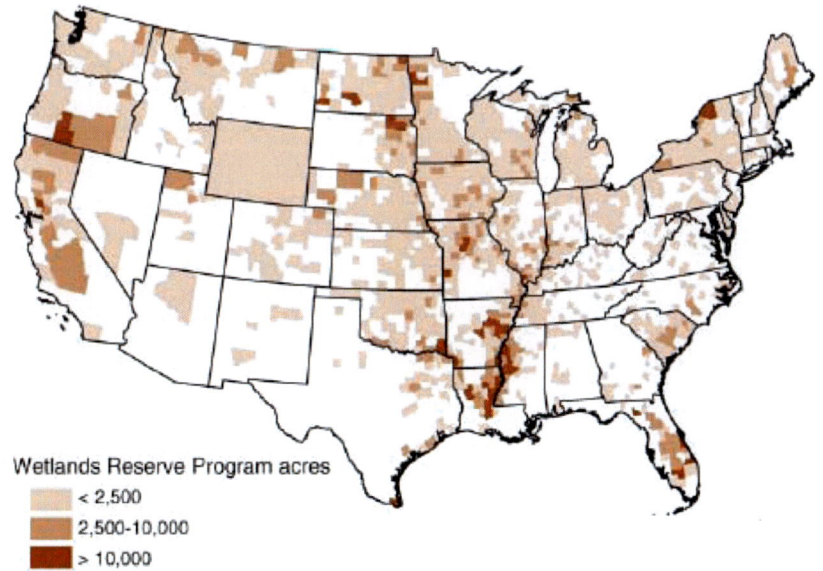

Source: USDA, ERS analysis of Wetlands Reserve Program contact data.

Figure 4.6. Wetlands Reserve Program acreage be county through 2005.

---

[5] ERS analyzed TNC project data. The contract data do not identify the wetland acreage involved, whether the projects involved land or easement purchases, or whether the projects involve existing or restored wetlands.

Wetland Mitigation Markets

Producers have some opportunities to sell wetland services directly to consumers, but functioning markets are rare. When able to control access to a wetland, landowners may have opportunities to sell rights to hunt, fish, and view wildlife. Limited data suggest that some producers are marketing fishing, hunting, and wildlife viewing on wetlands, but the extent is not known.

A better opportunity for producers to sell wetland services may exist in the offset markets created by Section 404 of the Clean Water Act. As in the case of water quality trading, a regulation is used to create demand for a private good (mitigation credits) that is closely linked to a public good (wetland environmental services).

The Act creates demand by requiring that any loss in wetland services be offset by a new or improved wetland that offers similar services (known as mitigation). Anyone wishing to drain or fill a wetland must first take all appropriate and practicable steps to avoid and then minimize harmful effects to wetland environmental services (U.S. Government Accountability Office, 2005). To offset any subsequent impacts, the firm or individual can either create wetland offsets (or credits) or purchase wetland credits from a mitigation bank.

The number of credits needed to mitigate lost wetland services is determined by a Mitigation Bank Review Team (MBRT), chaired by the U.S. Army Corps of Engineers (USACE). Besides the bank sponsor and wetland developer, participants typically include representatives from the EPA, Fish and Wildlife Service, National Marine Fisheries Service, and USDA's NRCS, as appropriate. Other Federal, State, and local governmental regulatory and resource agencies may participate, as well as tribal and other entities. Proposals must also be open to public review and comment.

Under Section 404, MBRT is to base its estimates of the number of credits a bank has available at a given time on the observed level of services and not on expectations of future additional services. That is, credits must be created before being sold. In theory, transactions involve no loss in wetland environmental services. The bank sponsor is responsible for creating, operating, and managing the bank; preparing and distributing monitoring reports, conducting compliance inspections of the mitigation, and securing funds for the long-term operation and maintenance of the bank (U.S. EPA, Office of Water, 1995b).

Banks may be sited on public or private lands (wildlife management areas, national or State forests, public parks, etc.). Federally funded wetland conservation projects undertaken via separate authority and for other purposes, such as the WRP, cannot be used in banking arrangements.

Banks must be located in an area (e.g., watershed, county) where it can reasonably be expected to provide comparable environmental services to offset the impact of wetland drainage. Data suggest that wetlands have been drained and mitigation banks created in both urban and nonurban counties (figure 4.7). Urban development pressure is a commonly cited reason for wetland loss, but clearly nonurban factors, such as highway construction and expansion, play a role.

The use of mitigation banks has increased steadily. In the 1980s and early 1990s, most wetland offsets were done through onsite restoration by developers. Few mitigation banking permits were approved. But in the mid-1990s, the number of bank approvals increased substantially (figure 4.8). Over the past decade, 30-50 mitigation banks have been approved annually. In total, over 600 mitigation banks have been approved or are under consideration for approval. Thirty-four States have at least one mitigation bank, but 80 percent are concentrated in 10 States (table 4.3).

[1] PIZA codes represent levels of urban development and are defined in USDA, ERS, 2005, http//ers.usda.gov/Data/PopulationInteractionZones/discussion.htm.

Figure 4.7. The distribution of mitigation banks and urban areas.

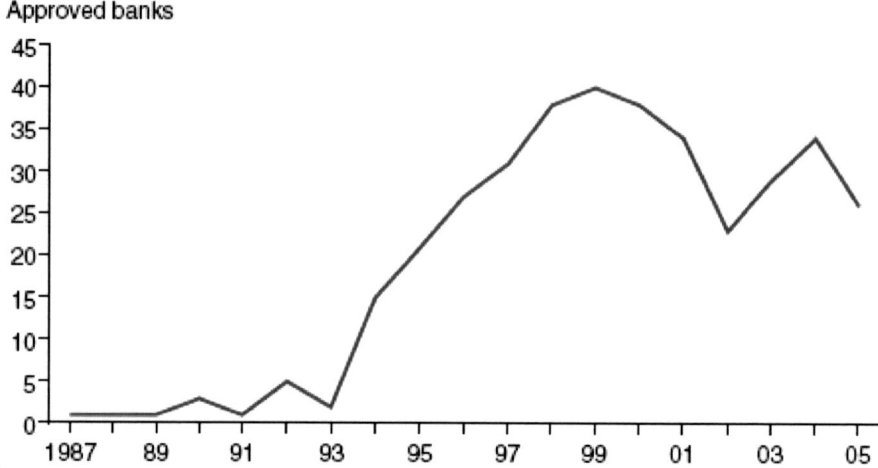

[1] Approval dates were not available for approximately half of the observations.
Source: USDA, ERS analysis of Environmental Law Institute mitigation banking data.

Figure 4.8. Mitigation banks approved 1.

Table 4.3. Number of approved mitigation banks by State

Rank	State	Banks	Rank	State	Banks
		Number			Number
1	Louisiana	100	18	Alabama	8
2	Georgia	74	19	Arkansas	8
3	California	61	20	Tennessee	8
4	Florida	55	21	Utah	7
5	Virginia	47	22	Idaho	5
6	Illinois	40	23	Indiana	5
7	Texas	22	24	Kentucky	5
8	Oregon	19	25	New York	5
9	North Carolina	16	26	Iowa	4
10	South Carolina	15	27	Nebraska	4
11	Colorado	14	28	Michigan	3
12	Mississippi	14	29	Delaware	2
13	Ohio	13	30	Maryland	2
14	Wisconsin	12	31	Kansas	1
15	Missouri	11	32	Oklahoma	1
16	New Jersey	9	33	South Dakota	1
17	Washington	9	34	West Virginia	1

Source: ERS analysis of U.S. Army Corps of Engineers data assembled by Environmental Law Institute.

Farmland owners should be in a good position to supply mitigation services. Nearly 60 percent of all mitigation counties have agricultural lands that were once wetlands. Prior wetland acreage is not a necessity—wetlands can be and are created on lands that have not previously been wetlands. But wetland restoration tends to be less costly on converted wetland acreage because soil type, topology, and other factors are favorable to wetland development. However, evidence suggests that agricultural landowners have played a small role in mitigation markets, although they may have sold lands to mitigation bank owners. In 2004, the first mitigation bank owned by a farmland owner was approved (USDA, NRCS, 2004a). Agricultural producers have not played a more direct role in wetland mitigation markets for a number of reasons.

Issues in Supply

Producers considering whether to become a mitigation banker could be influenced by several factors. Success of mitigation depends on, among other things, the permanence of the compensatory services. Mitigation permits must include a mechanism that guarantees long-term support of the wetland services. Bank sponsors are responsible for long-term costs of monitoring the wetlands, reporting banks' compliance, and maintaining the wetlands to ensure that banks' operations continue to provide the agreed-upon level of wetland services.

Uncertainty imposes an additional cost on suppliers. An entity wishing to produce and sell mitigation credits will not know the level of credits that a wetland will provide until after an MBRT's evaluation. Being uncertain of the credits a wetland may produce makes estimating income potential diffi - cult and may discourage investment in mitigation projects on the farm.

The long lag time between a wetland's restoration, the recovery of the wetland's environmental services, and the approval to sell credits can be a major issue (Shabman and Scodari, 2005). The loss in output from the land used to produce the bank and capital construction costs are certain, whereas income from the sale of credits is uncertain. Individual producers may not be comfortable taking on such a risk. Furthermore, the mitigation banker faces the risk of rule changes in the CWA or other legislation that might reduce or eliminate demand.

Issues in Demand

We assume that farmers would enter the mitigation market with the goal of maximizing profits. However, not all mitigation bankers have that goal.

Nearly 20 percent of all mitigation banking credits are supplied by nongovernmental organizations (NGO), that may be willing to absorb economic losses in exchange for increased wetland services. Farmers, therefore, may be at a competitive disadvantage and face a reduced demand for their wetland services in areas where NGOs are actively supplying wetland services (Shabman and Scodari, 2004). In addition, approximately 19 percent of all mitigation banking credits are supplied by government agencies, such as USDA and the Fish and Wildlife Service. The objectives of public agencies may or may not be profit maximization. In some cases, agencies have created more wetland credits than they need to mitigate their own actions, and have offered the excess in mitigation markets at reduced prices (Wilkinson and Thompson, 2006).

Demand for mitigation credits from producers may also be lost to in-lieu-fee mitigation. In-lieu-fee allows credits to be sold or accepted as offsets before being created, which eliminates many transaction costs and uncertainties faced by mitigation bankers and enables credits to be sold at a lower price than from a traditional mitigation bank. Only government agencies and NGOs are allowed to be in-lieu-fee bankers. Proposed changes to mitigation regulations could eliminate the cost advantage given in-lieu-fee banks (Kenny, 2007).

Balancing Supply and Demand

The success of the mitigation system depends on regulators' and arbitrators' abilities to recognize the quantity of services lost through development and provided by a mitigation bank. Participants in the MBRTs—mitigation bankers, developers, public agencies, NGOs, local communities—negotiate an agreement on the value of services lost and gained (U.S. Government Accountability Office, 2005). When participants have different goals, as is often the case (i.e., profit maximization versus ensuring protection of environmental services), negotiating trades can be time consuming and costly.

The Competitiveness of Mitigation Markets

We are interested in the extent to which producers might be able to benefit by participating in the wetland mitigation market. Data are too limited to allow us to estimate supply functions, but the available data allow us to compare wetland restoration costs. Based on WRP data from 1995 through 2007, county-level wetland restoration costs averaged $73-$525 per acre across counties with mitigation banks, with a maximum of about $2,500 per acre. Conversely, restoration costs of mitigation banks, in most cases, exceeded $5,000 per acre and, in some cases, exceeded $125,000. Assuming that the mitigation banks were successful financially, agricultural producers in those same counties would have also benefited if they had established mitigation banks on their land. While our analysis does not explain why we see such a difference, it gives us reason to believe that farmland owners may have a competitive advantage in wetland restoration.

The likelihood that a county will have a new mitigation bank can be estimated from historical data and measures of land characteristics. We estimated a probability model to predict the likelihood that a county will have at least one mitigation bank created in the future, given current development pressures. Factors that are likely to have an impact on the likelihood of a bank being developed include urban development pressure, wetland acreage (together with urban development, the source of demand), and total agricultural land (the likely source of supply). (For details of the model and the analysis, see Appendix: Predicting the Location of New Mitigation Banks.)

Based on the locations of current mitigation projects and the results of the model, 326 counties are predicted to be the most likely to see new mitigation banks in the near future (270 counties currently with mitigation banks and 56 additional counties with high development pressure and available wetlands). The likelihood for future mitigation projects is greatest in the coastal and Gulf States and parts of the Corn Belt, Lake States, and Mississippi Delta regions (fi g. 4.9). These counties contain 260,000 farms (12 percent of all farms). Farmers in these counties with the appropriate soils would have the greatest opportunity to create mitigation banks for the purpose of selling wetland services to developers. However, farmers may continue to decide not to accept the risk of becoming a mitigation banker but instead sell or lease land to mitigation banks.

MARKET INCENTIVES FOR WILDLIFE

Hunting is a popular recreation activity in the United States. Private lands are an important source of hunting opportunities. While wildlife residing on the land is a public good, the right to hunt on private lands is a private good controlled by landowners, one that can be sold to hunters willing to pay a fee. While some producers market hunting opportunities on their land, most do not. Thus, producers may have substantial opportunities to increase fee hunting, which could increase both producers' income streams and opportunities available to hunters. What's more, any increase in fee hunting may provide an economic incentive to producers to improve wildlife habitat that benefi ts both game and nongame species. In this case study, we identify factors that hinder the supply of and demand for fee hunting. We also consider the implications of using the

Conservation Reserve Program —a conservation program that pays farmers to retire land—to promote habitat improvement as well as access to hunting areas.

Background

Hunting in the United States is shaped by two fundamentals: (1) wildlife is owned in common by all citizens, and (2) most of the Nation's wildlife habitat is on private land (Benson, Shelton, and Steinbach 1999).6 Due to the dominance of private land ownership, Federal and State governments cannot exercise effective responsibility for wildlife management without productive collaboration with private land managers (Benson, 2001b; Conover, 1998).

Although wildlife is a public resource and individuals cannot claim ownership over it, private property access rights give landowners de facto control over wildlife residing on their land (Butler et al., 2005).

Reflecting this pattern of land ownership, the U.S. Fish and Wildlife Services' 2001 Fishing, Hunting, and Wildlife Associated Recreation survey (FHWAR2001) found that almost 75 percent of hunting days occurred on private land, 57 percent of all hunters hunted only on private lands, and nearly two-thirds hunted at least part of the time on private land.

Thus, private provision of the "hunting" environmental service is common. However, most of this provision does not rely on markets, where access is controlled by price. A 1993 national survey indicated that, while 77 percent of farmers allowed hunting, only 5 percent charged a fee (Conover, 1998). Several State studies report similar results.[7] Farm survey data from USDA indicate that only 1 to 2.5 percent of farms received income from recreation activities each year from 2000 to 2005 (USDA, ERS and National Agricultural Statistics Service).

While hunting on private lands is common, in recent years, many observers perceive that gaining access to private land for hunting has become more diffi cult (Larson, 2006; Bihrle, 2003). This observation is suggested by national data indicating that participation in hunting has dropped about 7 percent between 1996 and 2001 (U.S. Department of Interior, Fish and Wildlife Service and U.S. Department of Commerce, Bureau of the Census, 2002). This reduction may be partially explained by diffi culties in obtaining access to land. For example, a

---

[6] As stated in chapter 2, in 2002, private farms accounted for 41 percent of all U.S. land.

[7] The level of fee hunting tends to be higher in regions with smaller proportions of public lands, such as the South and Plains States (Langner, 1987; Conover, 1998). Jones et al. (1999) reported that, in Mississippi, 11-14 percent of landowners charged a fee for hunting in 1996-98, with gross revenues from hunting averaging about $3,300 per landowner in 1997 and 1998.

more urbanized population is less likely to have personal connections to rural landowners from whom they can easily obtain hunting access. Similarly, liability and other concerns seem to have driven an increase in the fraction of land that is posted (for no hunting).[8]

Yet, given the perceived decreased in supply, why is fee hunting so uncommon? Hunter surveys have consistently found that at least half would be willing to pay for hunting access (Benson, Shelton, and Steinbach, 1999). So why has fee hunting not expanded to satisfy demand, especially since income from recreation activities can be substantial?[9] Average gross revenue from fee-based recreation activities ranged between $13,000 and $18,000 per farm offering these activities between 2000 and 2005 (USDA, ERS and National Agricultural Statistics Service).[10]

Supply Issues

Farmer decisions about whether to market wildlife, such as through selling hunting access, hinge on several factors. These factors include attitudes about wildlife and permitting access to hunters, expected economic returns, and personal opinions about the marketing of wildlife.

Many producers apparently value having wildlife on their lands (Conover, 1998). A survey of producers' perceptions about wildlife on their farms found that 51 percent purposely managed their farm for wildlife. However, wildlife can also be seen as a problem to some farmers. The same survey found that 80 percent of the surveyed farmers experience wildlife-caused damage on their farms, and 53 percent stated that damage exceeded their tolerance levels. About a quarter indicated that wildlife damages reduced their willingness to enhance wildlife habitat. Farmers experiencing damage were more willing to allow hunting on their land, probably as a means of reducing wildlife damage, but this willingness to allow hunting does not mean producers would be willing to encourage wildlife by investing in habitat improvement.

---

[8] For example, Benson (2001) reports that 43 percent of State wildlife managers reported a decrease in hunting access between 1985 and 1994, whereas 8 percent reported an increase. In North Dakota, between 1992 and 2001, the share of landowners who posted their land increased from about 61 percent to over 68 percent (Bihrle, 2003).

[9] Some farms obtain more income from their hunting operations than from the crops they produce (Benson, Shelton, and Steinbach, 1999).

[10] These activities include hunting, fishing, petting zoos, tours, and onfarm rodeos.

One obstacle that limits greater use of markets may be the asymmetric distribution of costs and benefits. Wildlife does not respect property boundaries, which can limit the incentives for landowners to invest in habitat enhancements because some of the return will accrue to owners of adjacent parcels or even to other States (migratory waterfowl) (Lewandrowski and Ingram, 2001).

A fee hunting enterprise is not without cost. Setting up a fee hunting enterprise involves the time and expense of advertising, handling contracts, and addressing liability concerns on the farm. The latter is particularly important. The property must be inspected for abandoned wells, fences, dead trees, and other potential hazards that could lead to a liability suit if a hunter was injured. Also, a certain amount of compromise is necessary between the production of agricultural commodities and wildlife-related recreation in land management decisions in order to optimize income on all the land on the farm (Pierce, 1997). Management practices that can enhance game populations include brush control, planting perennial grasses, tillage practices, choice of crops, crop harvest, weed control, haying, grazing methods, stocking rates, fencing, and fertilization. Knowing how these practices affect game on the farm is critical to efficient management, and many landowners do not have the training to be effective wildlife managers (Butler et al., 2005). The higher the quality of the hunting or wildlife-viewing experience a farmer can offer, the greater the fee that can be charged.

Fee hunting does not always mean improved wildlife habitat. Many landowners who charge a fee are not increasing the provision of environmental services by managing their land for wildlife (Butler et al. 2005; Wiggers and Rootes, 1987; Benson, 2001a; Jones et al., 1999). In Mississippi, Jones et al. found that only 19 percent of farmers offering fee hunting actively managed their lands for wildlife.

Demand Issues

One of the largest issues in potential demand for fee hunting on private land is the belief that access to hunting areas should not be restricted by price. Many hunters and even landowners dislike the concept of fee hunting. For example, a North Dakota survey reports that over 50 percent of North Dakota farmers and over 60 percent of North Dakota hunters were "philosophically opposed to charging hunters for access" (Bihrle, 2003). A number of States actively promote open-access programs to counter fee hunting. In Washington, for example, a State program to open more private lands to hunters was instituted to "...combat the proliferation of fee hunting on private land...." (Washington Department of Fish and Wildlife, 2004). These programs pay a small fee, averaging about $5 per acre to participating landowners. In exchange, hunters are granted free walk-in access to the lands during hunting season without the need to obtain personal permission from the landowner. For most of these programs, the State also publishes (in print or on-line) land atlases that list all lands in the program. In several programs, participating landowners are covered under State liability insurance. Although such programs may compete with landowners who wish to charge a fee for access, they are likely to provide a "lower quality" recreational experience than is typical on fee hunting operations, which offer a wider range of services to hunters (Butler et al., 2005).

A Policy Simulation

Overall, the use of market mechanisms to provide wildlife-related environmental services, while not unusual, is not widespread. Obstacles include the public-goods nature of wildlife, inadequate education of potential private benefits from developing a fee hunting business, the complications of operating a hunting and farming business on the same land, and the transaction costs of bringing potential demanders and suppliers together. In this section, we consider how a Federal conservation program might be used to increase the willingness of producers to invest in improved habitat and supply hunting opportunities on their land for a financial gain—in other words, using an existing program to kick-start the market.

USDA's Conservation Reserve Program (CRP) is an example of a payment-for-environmental-services program. Established by the Food Security Act of 1985, the program uses contracts with agricultural producers and landowners to retire over 34 million acres of highly erodible and environmentally sensitive cropland and pasture from production for 10-15 years. When first started, the CRP's primary goal was soil conservation. However, it has evolved beyond soil conservation, with greater weight given to wildlife habitat and air and water quality. The CRP has successfully provided a variety of environmental services, including significant reductions in soil erosion, hence, cleaner waterways, and large increases in wildlife populations. However, the question remains whether market mechanisms can be harnessed to further increase the benefits of land retirement and the quality of wildlife habitat.

The CRP could provide for more improved habitat than fee hunting alone, bring together buyers and sellers, and provide economic opportunities for landowners in areas where there are cultural objections to fee hunting. A number of States with walk-in access hunting programs specifi cally target land enrolled in CRP (table 4.4). The use of CRP to promote both wildlife habitat enhancement and hunting would also have implications for the distribution and rental rates of enrolled acres. Enrollment decisions by USDA could favor landowners who allow public access to their lands. To gauge how an aggressive policy of using the CRP to market hunting services would affect enrollment, several scenarios are examined using a simulation model combined with a measure of potential hunting demand.

The CRP uses an Environmental Benefi ts Index (EBI) to determine what lands to accept from among all farmer offers (USDA, ERS, 2007b). The EBI has a wildlife component that could be modifi ed to reward offers that permit public access, even if the landowner charges a fee. Such a modifi cation could substantially increase hunting access in some States. However, it might have little net impact in regions where alternatives (public land or an active market for leases) are available.

We used the USDA Farm Service Agency's Likely To Bid (LTB) model to predict what lands would be enrolled in the CRP if landowners were provided an additional incentive to be more open to selling the "hunting" environmental service (say, through a federally operated program). Data from the 2001 Fishing, Hunting, and Wildlife Associated Recreation Survey (FHWAR2001) and 2000 National Survey of Recreation and the Environment (NSRE2000) are used to identify the potential demand for hunting:

- The NSRE2000 is a midsized dataset (several thousand respondents) that asks questions about participation in wildlife-related activities. About 1,600 respondents provided a distance (from their residence) to the location they visited on a wildlife-related trip and a direction (i.e., 120 miles to the north).
- The FHWAR2001 is a large dataset (about 25,000 respondents) that asks extensive questions about hunting, including States in which the respondents hunted.

The FHWAR200 1 provides accurate measures of how many hunting trips were made to each State. The NSRE2000 provides a relative measure of where hunters went within each State. Combining the two provides an estimate of total trips hunting, a measure of "hunting pressure," for all U.S. counties. For this simulation, we ranked all counties by this "hunting pressure" measure and classified the upper 50 percent as "hunting counties." Figure 4.10 displays the results of combining these two datasets, with counties in green being "hunting counties" and urbanized areas in orange.[11]

Table 4.4. States with walk-in hunting-access programs that enroll CRP acreage

State	CRP acreage in program	2005 CRP acreage in State	Notes
Acres		*Million acres*	
Colorado	135,000	2.2	Total program size is 160,000 acres
Kansas	534,000	2.8	Total program size is about 1 million acres http://www.kdwp.state.ks.us/
Nebraska	180,000	1.2	Complements 250,00 acres of Park and Game Commis sion land http://www.ngpc.state.ne.us/hunting/programs/crp/crp.asp
South Dakota	About 333,000	1.5	About 1,000,000-acre program (private communication, SD Di vision of Wildlife, Bill Smith)
North Dakota	About 180,000	3.3	About 425,000 acres in ND Private Land Initiative. http://www.nodakoutdoors.com/valleyoutdoors10.php

CRP = Conservation Reserve Program.
Source: As noted and from Helland (2006). Several other States have walk-in hunting-access programs that do not use signifi cant CRP acreage. Payments to landowners may depend on the quality of the wildlife habitat and on habitat-improving practices installed by the landowner.

We specify a baseline scenario under current policy and compare this baseline to several alternatives. In all the alternatives, only acreage in "hunting counties"

---

[11] The NSRE has about 100 observations that could be classifi ed as "hunting trips to CRP-like lands." Thus, in order to get a reasonable national coverage, all observations (for all wildlife-related trips) were used, which may introduce bias because hunting trips are probably to locales that systematically differ from trips for other wildlife- related recreation. Nevertheless, the NSRE does capture the distribution of population and does relate to wildlife- associated recreation.

responds to the scenarios' postulated changes.12 The scenarios examined are as follows:

1. Where demand is high, farmers recognize their potential to sell hunting leases. They lower their offer prices (the amount they ask for to enroll their land in the CRP) in response to potential income from retired land. This scenario assumes a reduction of $5 per acre in the offer price ($5 per acre is an upper-end value for several of the State "walk-in" programs).
2. Where demand is high, the government successfully encourages applicants to maximize the N1 "wildlife points" in the EBI. In practice, this might be achieved by fully cost-sharing wildlife-enhancing practices (rather than the standard 50-percent cost share).
3. Combination of 1 and 2: Farmers lower offer prices, and the government subsidizes wildlife practices.
4. Similar to 3, but landowners do not lower their bids, although they still assume they can earn $5 per acre from hunting leases. While this hunting lease income does not influence the EBI scores of submitted bids, it does increase the acreage offered to the program (since farmers will receive both government and hunter payments for their CRP land).

Table 4.5 summarizes the results of the policy experiments and of a baseline that uses the current CRP rules. The model is calibrated for 2006, so it does not reflect current high commodity prices. However, comparisons of scenarios to the baseline should be roughly accurate. Note that, in all scenarios, a 35-million-acre program is simulated.13 These results are best used as indicators of the range and types of changes to the CRP rather than of specific predictions.

The most general results are not surprising: The average bid decreases when expected, and the wildlife score (N1) increases. However, a few points are worth noting:

1. Acreage shifts from nonhunt counties to hunt counties can be substantial: In the fourth scenario, over 3 million acres shift, leading to about a 20-percent increase in hunt-county CRP acreage.[14]

---

[12] Demand for hunting is relatively low in "nonhunting counties"; thus, in these nonhunting counties, the alternatives assume no changes in the factors that influence which acres are offered to the CRP.

[13] The size of the CRP will be reduced to 32 million acres over the next several years, as specified in the Food, Conservation, and Energy Act of 2008.

[14] An extension to scenario 4, that classifies the Northern Plains States with "walk-in hunting access programs" (North Dakota, South Dakota, Kansas, and Nebraska) as "hunting pressure," yielded

2. The average bid does not decrease by $5 per acre in hunt counties, which is due to the heterogeneity of land types and the increased likelihood of accepting land as bid rates drop. Thus, decreasing the bid of a previously rejected "environmentally desirable but expensive" offer may result in its acceptance, even though its offer price may still be greater than average.
3. N1 scores increase by about 25 percent. Although occurring largely in hunt counties, wildlife scores also increase in nonhunt counties. The increase in average EBI scores causes marginal offers in nonhunt counties (those that often have relatively low N1 scores) to be dropped, increasing the overall average.
4. The fraction of landowners willing to make offers to the program increases when they are assumed not to lower their bids (scenario 4).

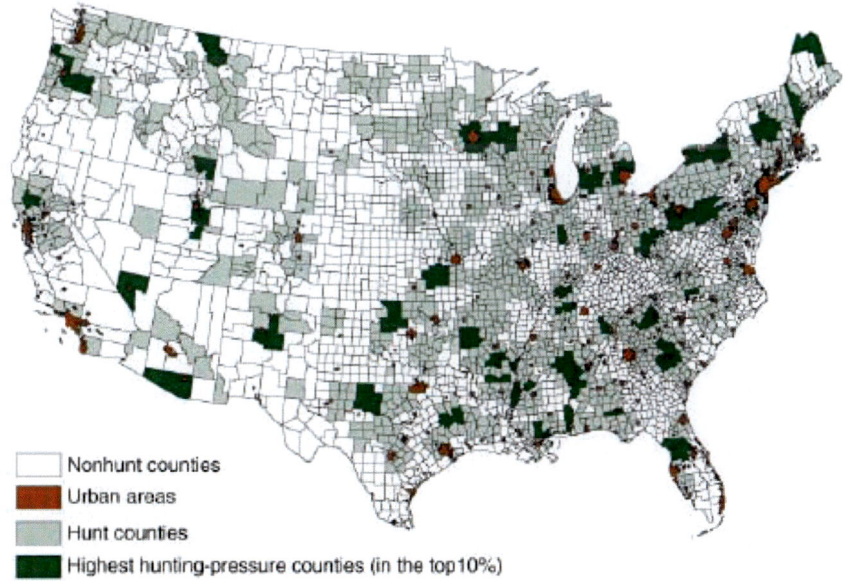

Source: USDA, ERS, using FHWAR2001 and NSRE2000 data.

Figure 4.10. Hunting pressure by county.

similar results, although the acreage in the Northern Plains increases substantially, largely at the expense of the Mountain, Southern Plains, and Corn Belt regions.

Table 4.5. General results of scenarios

Item	Baseline	Lowered bids	Increased wildlife points	Lowered bids and increased wildlife points	Increased income and increased wildlife points
Acreage enrolled (million acres):					
All counties	35.0	35.0	35.0	35.0	35.0
In nonhunt counties[1]	19.7	18.8	17.2	16.1	16.5
In hunt counties	15.3	16.2	17.8	18.9	18.5
Average bid of enrolled acres ($):					
Across all counties	24	23	25	23	26
In nonhunt counties	21	21	21	21	21
In hunt counties	28	25	28	25	30
Average N1 score of enrolled acres:[2]					
Across all counties	74	74	82	83	75
In nonhunt counties	74	74	75	75	90
In hunt counties	73	73	90	90	83

[1] "Hunt counties" are counties identified using the hunting-pressure index (all counties with a hunting-pressure index score greater than the median score).
[2] The maximum value of the N1 component of the Environmental Benefits Index is 100.

In addition to national impacts, the regional distribution of CRP land may change under different scenarios. Changes across the 10 USDA Farm Production Regions are summarized in figure 4.11. The most striking result is the acreage reduction (compared with the baseline) in the Northern Plains, Pacific, and Mountain States. These acres are reallocated to the other regions, especially the Corn Belt and Lake States, which are largely driven by the greater hunting pressure east of the Mississippi.

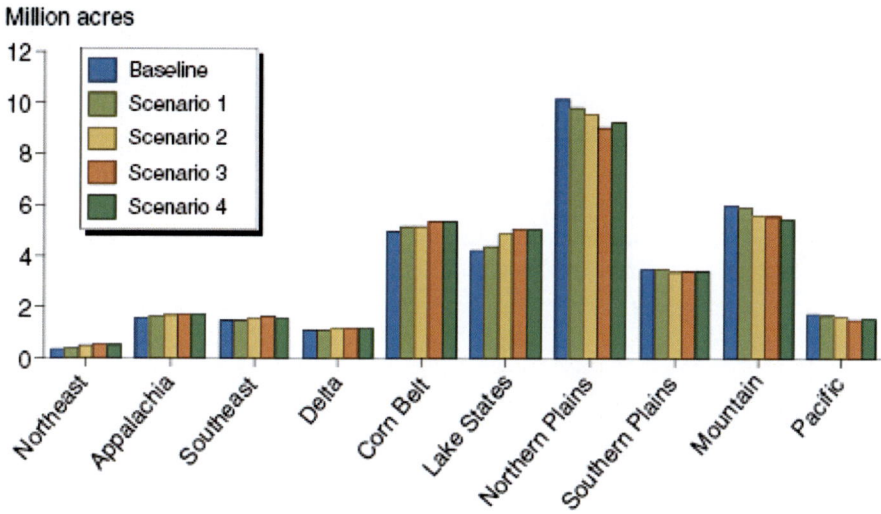

Scenario 1: Farmers reduce CRP offers by $5.
Scenario 2: Farmers maximize wildlife points in Environmental Benefits Index and receive full-cost share.
Scenario 3: Farmers reduce CRP offers by $5, maximize wildlife points, and receive full-cost share.
Scenario 4: Farmers do not lower CRP offers, maximize wildlife points, and receive full-cost share.
Notes: The scenario descriptions summarize several possible per acre changes. The possibility of leasing their CRP land to hunters may lead landowners to reduce their bids, to improve the wildlife habitat on their land, or to factor in lease income when deciding whether or not to offer their land to the program.
Source: USDA, ERS.

Figure 4.11. Enrollment in Conservation Reserve Program (CRP) by region for fee hunting scenarios.

Scenarios where fee hunting opportunities cause landowners to reduce their offers (hence increasing the likelihood of an offer acceptance) yield similar results to scenarios where offers are not reduced (hence increasing the likelihood of an offer being made). There are a few differences. For example, in the Northeast, acreage is slightly higher in the "reduce-the-bid" scenario than in the "do-not-reduce-bid" scenario. Conversely, in the Northern Plains, the opposite is observed.

Implications

Recreational hunting primarily occurs on private lands. While most of this access is through informal mechanisms, landowners have a long history of charging willing hunters to access their land. However, for a variety of reasons, the marketing of the "hunting" environmental service is still relatively small. And although some of these reasons (such as landowner reluctance to give strangers with guns access to their property) are unlikely to change, others (such as the difficulty of connecting landowners to hunters) may be amenable to institutional solutions. However, even with greater farmer participation, evidence suggests that fee hunting does not always lead to improved wildlife management, which is a major reason for promoting the creation of markets.

One institutional solution for improving wildlife habitat is via government policy vis-à-vis agricultural conservation programs, such as the CRP. While only suggestive, the alternative scenarios for coordinating the CRP with hunting access indicate that such programs could reduce CRP costs (with a 10-percent reduction in offer price in some scenarios) and increase the quality of wildlife habitat (with a 25-percent increase in one measure of wildlife habitat in some scenarios). These scenarios also suggest that CRP acreage may shift toward more populated areas of the country (where there are more hunters). An indirect benefit of this could be increased values from the provision of other environmental services, such as open space and water quality.

Overall, current trends suggest greater restriction on casual access to hunting lands, with continued urbanization further weakening the link between nonrural hunters and rural landowners. Thus, the prospects of using private provisions of hunting services are likely to increase.

"USDA ORGANIC" AND OTHER ECO-LABELS IN AGRICULTURE

One way that a farmer could benefit financially by providing an environmental service is to link the provision of the service to the sale of a private good. Eco-labeling is a way of informing consumers of the process used to produce the private good and, concurrently, its impact on environmental services.

Consumers who care about environmental services may be willing to pay a higher price for products produced in a way that provides those services.

Starting with the organic label in the 1950s, eco-labels have been used to tout reduced pesticide use, wildlife protection, and other environmental services tied to specifi c agricultural production systems. Food that has an organic or other eco-label is fundamentally a "credence good"—it cannot be distinguished visually from conventional food—and consumers must rely on labels and other advertising tools for product information. Many consumers associate enhanced food safety and nutrition, environmental protection, and other qualities with eco-labels. We examine experience with the organic label for lessons on how this approach could be expanded to a wider set of environmental goals.

National Organic Standards Define an Ecological Production System

The organic label is the most prominent eco-label in the United States, refl ecting decades of private-sector development and subsequent initiation of a government regulatory program. Congress passed the Organic Foods Production Act of 1990 (OFPA) to establish national standards for organically produced commodities in order to facilitate domestic marketing of organically produced fresh and processed food and to assure consumers that such products meet consistent, uniform standards.

The program establishes: (1) national production and handling standards for organically produced products, including a national list of substances that can and cannot be used; (2) certifi cation requirements for organic growers; (3) a national-level accreditation program for State and private entities, which must be accredited as certifying agents under the USDA national standards for organic certifi ers; (4) requirements for labeling products as organic and containing organic ingredients; and (5) civil penalties for violations of these regulations.

In setting the soil fertility and crop nutrient management practice standard, USDA requires the producer to use practices that maintain or improve the physical, chemical, and biological condition of soil and minimize soil erosion. The producer is required to manage crop nutrients and soil fertility through rotations, cover crops, and the application of plant and animal materials and is required to manage plant and animal materials to maintain or improve soil organic matter content in a manner that does not contribute to contamination of crops, soil, or water by plant nutrients, pathogenic organisms, heavy metals, or residues of prohibited substances.

Environmental benefi ts that can be attributed to organic production systems include the following:

- *Reduced pesticide residues in water and food.* Organic production systems virtually eliminate synthetic pesticide use, and reducing pesticide use has been an ongoing U.S. public health goal as scientists continue to document their unintentional effects on nontarget species, including humans.
- *Reduced nutrient pollution, improved soil tilth, soil organic matter, and productivity, and lower energy use.* A number of studies have documented these environmental improvements in comparing organic farming systems with conventional systems (USDA Study Team on Organic Farming, 1980; Smolik et al., 1993; Mäder et al., 2002; Marriott and Wander, 2006).
- *Carbon sequestration.* Soils in organic farming systems (which use cover crops, crop rotation, fallowing, and animal and green manures) may also sequester as much carbon as soils under other carbon sequestration strategies and could help reduce global warming (Lal et al., 1998; Drinkwater et al., 1998). (See "Issues in Supply of Carbon Sequestration" for more detail.)
- *Enhanced biodiversity.* A number of studies have found that organic farming practices enhance the biodiversity found in organic fields compared with conventional fields (Mäder et al., 2002; Altieri, 1999) as well as improving biodiversity in field margins (Soil Association, 2000).

Issues in Supply: Major Farm Sectors Lag in Adopting Organic Systems

U.S. farmland under organic management has grown steadily for the last decade as farmers strive to meet consumer demand in both local and national markets. U.S. certified organic crop acreage more than doubled between 1992 and 1997 and doubled again between 1997 and 2005 (USDA, Economic Research Service, 2007c). Organic fruit and vegetable crop acreage, along with acreage used for hay and silage crops, expanded steadily between 1997 and 2005. However, most of the acreage increase for organic grain and oilseed crops took place early in this period, and organic soybean acreage has declined substantially since 2001.

California had more certified operations than any other State, with just over 1,900 operations in 2005, up 20 percent from the previous year. Wisconsin, Washington, Iowa, Minnesota, New York, Vermont, Oregon, Pennsylvania, and Maine rounded out the top 10. Many of these States have a high proportion of farms with fruits and vegetables and other specialty crops. Also, some of these States, particularly in the Northeast, have relatively little cropland but a large concentration of market gardeners.

While adoption of organic farming systems showed strong gains between 1992 and 2005 and the adoption rate remains high, the overall adoption level is still low: Only about 0.5 percent of all U.S. cropland and 0.5 percent of all U.S. pasture were certified organic in 2005 (fig. 4.12). About 8,500 operations are certified organic (out of over 2 million farms).

One of the biggest obstacles to adoption is the cost of converting a farm to organic production. The transition from conventional production systems to organic systems typically involves high managerial costs and risks (Oberholtzer, Dimitri, and Greene, 2005). Production costs may be higher because of more intensive use of labor, use of substitutes for synthetic chemicals, longer crop rotations for disease and pest control, reduced yields, and increased recordkeeping.

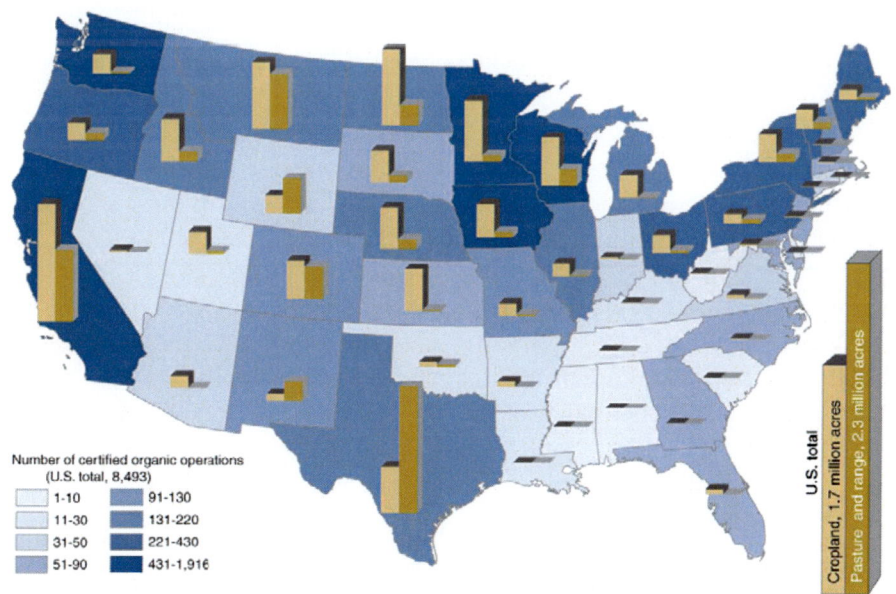

Source: USDA, ERS, based information from USDA-accreditied State and Private organic certifiers.

Figure 4.12. U.S. certified organic acreage and operations, 2005.

Another issue is access to production and market information. The infrastructure for extension, marketing, handling, and transport is far less developed than for conventional production systems (Lohr and Salomonsson, 2000). A lack of publicly funded organic farm advisors and relatively little government-funded research on organic production and marketing systems has hindered adoption in the recent past (Lipson, 1997). Most organic information is disseminated by farmers and private organizations (Lohr and Salomonsson, 2000).

Issues in Demand

Farmers, food processors, and other businesses that produce and handle organically grown food have a financial incentive to advertise that information because consumers have been willing to pay a price premium for these goods. Academic research studies in the 1980s and early 1990s found that consumers were purchasing organic products in response to environmental concerns, such as the impacts of pesticide use on the environment, groundwater, wildlife, and agricultural workers, as well as personal safety concerns (Bruhn et al., 1991; Weaver, Evans, and Luloff, 1992; Cuperus et al., 1996; Goldman and Clancy, 1991; Davies, Titterington, and Cochrane, 1995; Morgan, Barbour, and Greene, 1990).

Although eco-labels enable consumers who value environmental services to pay for the services through the purchase of certain goods, environmental services are still public goods, which presents an opportunity for some of those who value these services to free ride. To the extent that free riding occurs, the prices that organic farmers receive do not refl ect the full value that society places on the environmental services provided by this approach to farming.

A negative feature of a proliferation of eco-labels and other labels related to such issues as social justice is that label effectiveness may diminish because multiple, competing label claims may cause consumer confusion (U.S. EPA, 1998). On the other hand, many consumers may be savvy enough to see the differences between unregulated labeling terms that indicate the use of some alternative production practices and a government-regulated, fully defi ned, independently certifi ed product label like "USDA organic." Also, some of these labels are complementary, not competitive. Organic certifying entities, both State and private, already certify producers and processors to a number of other standards—including food safety standards and international organic standards that already incorporate a social justice component. A product might easily carry both an organic label, denoting the ecologically based production system used, and a locally grown logo, denoting the number of food miles to deliver the product to the consumer.

Table 4.6. Eco-labels used in U.S. food and agriculture sectors

Label	Private standards	Government standards	Program certification	Certified operations, 2005/06 Number	Farmland 2005/06 Acres	Retail sales 2006 $ million
Process-based						
USDA organic (organic production and food processing systems)	International (IFOAM; Codex)	USDA-AMS, Federal Register, National Organic Program, final rule, Dec. 21, 2000, pp. 80548-80684	Private-1971 State-1980 Federal-2000	50 States; Farmers—8,493 Processors—3,000	Cropland—1,723,271 Pasture—2,331,158	16,000
Healthy Grown (alternative pest management)	World Wildlife Fund-protected harvest	No	1998	Wisconsin; Farmers—11	Cropland—5,823	—
Salmon Safe (alternative production and salmon habitat restoration practices)	Salmon safe	No	1997	Northwest (4 States); Farmers—39	Cropland and pasture—30,000	—
Pure Catskills (alternative production practices)	Watershed Agricultural Council	No	Promotion only	New York; Farmers—102	Cropland—	—
Responsible Choice (alternative apple production, packing, and shipping practices)	Stemult Growers, Inc. (1989)	—	1989	Farmers—250 Processor—1	Cropland—	—
Product-based						
Natural (minimal processing, no artificial ingredients, additives, or coloring)	—	Definition, but no standards (FTC, 1970s; USDA, 1982)	Promotion only	—	—	5,140
Location-based						
State logos (such as Jersey Fresh, Minnesota Grown, Pride of New York, and Virginia's Finest)	—	State departments of agriculture	Promotion only (first logo, 1983)	44 States	—	—
Food miles (CO_2 emissions)	Iowa State University Leopold Center (pilot)	No	—	—	—	—
Social Justice						
Food Alliance Certified (standards for working conditions and alternative production practices)	Food Alliance	No	1998	10 States; Farmers—159	Cropland—156,001 Pasture—4,148,467	82
Just Organic (farmers' rights, farm workers' rights, fair trade and indigenous peoples' rights)	Florida Certified Organic Growers and Consumers (pilot)	No	—	—	—	—

IFOAM=International Federation of Organic Agriculture Movements; AMS=Agricultural Marketing Service; FTC=Federal Trade Commission.

Sources: USDA-ERS, www.ers.usda.gov/briefing/organic; Wyman, 2006; Food Alliance, www.foodalliance.org; and Saam, 2007.

Emerging Eco-labels

In addition to the organic label, a number of other eco-labeling programs have emerged in the food and agricultural sector for a broader group of farm-related characteristics (table 4.6). Some of these programs use private third-party certifi cation to enhance consumer confi dence, but none has a government regulatory program similar to the organic program.

Several process-based labels have emerged, which have a regional focus. In 1998, the World Wildlife Federation collaborated with another nonprofi t, "Protected Harvest," to initiate a label for potato farmers in Wisconsin that would reduce the use of some toxic pesticides and encourage other environmentally benefi cial production practices. About the same time, a nonprofi t in the Pacifi c Northwest developed a "Salmon Safe" label that recognizes the adoption of "ecologically sustainable agricultural practices that protect water quality and native salmon." This label encourages restoration of riparian habitat adjacent to fields, as well as improved cropping system practices. Fewer than 50 farmers were using these programs in 2005/06. About 100 growers in New York, using a variety of production systems, were using a "Pure Catskills" promotional label to indicate their participation in a watershed protection program.

Location-based labels may have had limited use in this country, but a "food miles" label may emerge as interest in reducing the energy costs and environmental impacts of food transportation increases (Leopold Center, 2003). States have been developing promotional logos to appeal to consumer interest in helping to protect their State's farmland from development since the early 1980s, and 44 States now have their own agricultural logo—i.e., "Jersey Fresh" and "Virginia's Finest." Many consumers may associate environmentally friendly production practices with local production and local labels. However, these labels address product freshness and the energy used in transportation during the food distribution process but not necessarily environmentally friendly production practices.

Summary

The organic label is the most important eco-label in the United States. It has benefi ted from consumer demand, a clearly defi ned set of standards, a strong certifi cation system, and a system of enforcement. However, the adoption of organic production systems is still fairly low. Obstacles to adoption by farmers include high managerial costs and risks of shifting to a new way of farming, limited awareness of organic farming systems, uncertainty over expected yields and returns, lack of marketing and infrastructure, and inability to capture marketing economies (Greene, 2001). These factors are likely to be issues for other types of eco-labels as well.

The proliferation of other local and national eco-labels for a variety of environmental services and labels for other causes may pose a challenge to consumers. Many of these labels do not come with the standards and certification of the organic label, raising the uncertainty of the label claims. Even if consumers are willing to pay a premium to support the supply of environmental services on farms, too much information may make deciding between competing goods difficult. However, careful development of new production standards and labeling regulations, along with consumer education, production research, and other policy initiatives, can mitigate consumer confusion and address the obstacles to adoption.

Even if price premiums for eco-labels can be maintained, however, the public-goods nature of environmental services, such as biodiversity and water quality, implies that they do not reflect the true social value of these services. Eco-labels alone do not provide a socially optimal level of environmental services.

Chapter 5

LESSONS LEARNED AND POTENTIAL ROLES FOR GOVERNMENT

Producers have opportunities to sell environmental services in a number of well-functioning markets (table 5.1). EPA is encouraging States to use the Clean Water Act's permit program to establish markets for pollution discharge allowances and to include agriculture in these markets. Producers can sell credits for greenhouse gas reductions on the Chicago Climate Exchange and in a growing number of retail carbon markets. Wetland mitigation markets are operating in many States, and the concept has been expanded to protect endangered species habitat. Fee hunting operations are commonplace in a few States and demonstrate that producers can earn substantial income that could be used to support wildlife habitat. Organic labeling is well established, and food labels are expanding to include information related to the provision of a wider set of public goods on farms.

Overall, however, farmer participation in these markets has been limited. Part of the reason is that many of the markets themselves are limited in scope. Experiences with these markets have also identified a number of impediments that limit producers' participation. Many of these impediments are unlikely to be overcome without direct involvement by government, including USDA. USDA has already identified some of the actions it can take to assist in the development of markets and to increase farmer participation (see box, "USDA Commitments to Markets for Environmental Services"). Economic theory and experience with the markets described in the case studies highlight a number of issues that are of primary importance in the successful creation of markets for environmental services.

Table 5.1. Summary of existing markets for environmental services and some important characteristics

Market	Water quality trading	Chicago Climate Exchange	Retail carbon market	Wetland mitigation banking	Organic labeling	Fee hunting
Environmental service	Water quality	Reductions in net greenhouse gas emissions	Reductions in net greenhouse gas emissions	Wetland services	Various (water quality, biodiversity, air quality)	Wildlife
Good traded	Discharge allowance	Carbon credit	Carbon credit	Qualified wetland acreage	Agricultural food, fiber, and other products	Access to land
Source of property right	Regulatory agency	CCX rules	Retail carbon provider	Regulatory agency	Private good	Private good
Source of demand	Regulatory discharge cap on point sources	Legally binding discharge cap on member firms	Private sentiment	Legally binding no-net loss rules	Private sentiment	Private sentiment
Standards?	Yes	Yes	No	Yes	Partial	No
Steps being taken to reduce uncertainty	Research on performance of conservation practices, flexible rules for point sources, verification, enforcement	Research on performance of conservation practices, verification	None	Research on measuring and verifying wetland services	Uniform national standards, mandatory certification, Federal enforcement	Research on improving habitat, outreach
Steps being taken to reduce transactions costs	Third-party aggregator, clearinghouse, outreach, models	Third-party aggregator, models, Voluntary Greenhouse Gas Reporting Registry	Online decision aids	Third-party arbitrators	Reduction in multi-ingredient certification disputes	Outreach, clearinghouse operated by State, liability coverage
Remaining impediments or issues	Producer reluctance, lack of binding caps, interactions with conservation programs	Lack of national binding cap, interactions with conservation programs	Lack of standards and verification	Up-front costs and market uncertainty, interactions with conservation programs	Information overload, free-riding on environmental benefits	Public sentiment, free-riding on wildlife services

CCX=Chicago Climate Exchange.

> ## USDA COMMITMENTS TO MARKETS FOR ENVIRONMENTAL SERVICES
>
> In 2006, USDA released a departmental regulation defi ning its policy on markets for environmental services. This policy stated that USDA would do the following:
>
> - Cooperate with other Federal, State, and local governments to establish a role for agriculture in environmental markets.
> - Find ways to make USDA policies and programs support producers wanting to participate in such markets.
> - Conduct research and develop tools for quantifying environmental impacts of farming practices.
>
> A partnership agreement between EPA and NRCS to collaborate on efforts to establish viable water quality trading markets was signed in 2007. A goal is to develop a pilot water quality trading project in the Chesapeake Bay watershed.
>
> The Food, Conservation, and Energy Act of 2008 contained a section in the Conservation Title outlining USDA's role in support for market-based conservation. The provision required the following:
>
> - The Secretary of Agriculture will establish technical guidelines for measuring environmental services from conservation and other land management activities, and priority will be given to developing guidelines for participation in carbon markets.
> - Guidelines will be established for a registry to collect, record, and maintain information on measured benefits.
> - Guidelines will be established for a process to verify that a farmer has implemented the conservation or land management activities reported in the registry.

ISSUE: PERFORMANCE OF MANAGEMENT PRACTICES

One of the biggest issues facing producers who wish to participate in markets for environmental services is uncertainty about the environmental performance of conservation practices, such as conservation tillage, riparian buffers, and nutrient management. In emissions trading and offset markets, uncertainty about the quantity of credits that can be supplied reduces demand for environmental services from agriculture. Markets often try to account for this uncertainty by requiring that a lost unit of wetland services or a point- source unit of pollution discharge be replaced or mitigated with two or more units of services (credits) from farms. This practice essentially increases the price of mitigation to buyers and reduces overall demand for farmer- produced credits.

Uncertainty of practice performance also affects the potential supply of environmental services. Uncertainty about the quality or quantity of the environmental services a farm can produce makes it difficult for producers to decide the long-term economic benefit of investing in a wetland mitigation bank, to make wildlife habitat improvements for a fee hunting business, to enter an emissions trading market, or to enter the organic market. Uncertainty about the impact of a new practice on crop yields can also affect a farmer's decision to implement a practice in order to enter a market. In the case of the Chicago Climate Exchange, lack of scientific evidence about a soil's ability to sequester carbon can prevent a farm from entering the market.

USDA can play a role in providing research on the effectiveness of different conservation practices for producing environmental services. USDA already provides farmers and ranchers with information on the impact of conservation practices on air, water, and wildlife habitat through sources like the NRCS Field Office Technical Guide. However, much more detailed information is needed to estimate the number of credits that might be produced for sale in emissions trading markets or the wetland services that can be sold by a mitigation bank.

USDA supports the development of tools and methods for quantifying how changes in farming practices affect environmental services (USDA, Natural Resources Conservation Service, 2006b). For example, the Nitrogen Trading Tool and GRACEnet can help reduce uncertainty in water quality and carbon trading markets, respectively.

Another broader effort is the Conservation Effects Assessment Project (CEAP). The goal of CEAP is to quantify the environmental benefits of conservation practices used by private landowners participating in USDA conservation programs. Field-level sampling, monitoring, and modeling are being used to estimate the impacts of conservation practices on water quality, wildlife, and soil quality. In addition, collaborative regional assessments are developing models for estimating environmental services from wetlands, including carbon storage, sediment, and nutrient reduction, flood water storage, wildlife habitat, and biological sustainability (USDA, NRCS, 2006a). CEAP also includes watershed assessment studies that are to provide a framework for evaluating and improving the performance of water quality assessment models. Such models are critical for estimating the equivalency of water quality credits that are produced in different parts of a watershed. Models that can predict the movement of chemicals carried in runoff with a degree of certainty sufficient to allow agricultural credits to be traded would make it easier for producers to participate in trading programs. Models would also allow uncertainty ratios (trading ratios that specifically reflect practice uncertainty) to be lowered, reducing the cost of agricultural credits and making them more attractive to point sources. In addition, research sponsored by the USDA Cooperative State Research, Education, and Extension Service is also addressing practice performance in a variety of settings, as well as supporting the development of assessment tools.

Uncertainty over the economic performance of practices implemented to produce environmental services can also be overcome through risk-management instruments, such as insurance (Zeuli and Skees, 2000). Private companies could provide such instruments, but government could also offer them if an active market for environmental services is an important conservation goal and private insurance is not available.

ISSUE: STANDARDS AND VERIFICATION

One of the requirements for a smoothly operating market is that the good being traded is of a consistent quality that is known to all. Organic agriculture and emissions trading markets have very specific standards for the services that are marketed, which is not the case for the retail carbon market and some of the newer eco-labels. Consumers may not know what they are buying or how the environmental services provided by one supplier differ from another. For example, what does "wildlife-friendly" agriculture really mean? What does it really take to eliminate the carbon footprint of an airline flight or a wedding? As long as labels and advertising are the only ways consumers have of discriminating between the ability of producers to provide environmental services, consumers are likely to be skeptical of suppliers' claims. Third-party certification is considered the only reliable way to signal product quality claims in organic markets (Cason and Gangadharan, 2002).

USDA is playing an important role in setting standards and providing certification for organic agriculture. Standards and certification provide the assurance to consumers that the claims on the label are believable and protect producers from dilution of price premiums due to less rigorous (and less costly) applications of organic standards. The department regulation outlining USDA's roles in "market-based stewardship" calls for USDA to cooperate with other Federal Departments and groups in developing accounting practices and procedures for quantifying environmental goods and services in other types of markets. Research on practice performance would help USDA contribute to such a role.

Verification that standards are being followed and that promised management practices are being implemented is a related issue. Many environmental services are not easily observed. Verification is based on the farming practices that have been implemented, and this often requires on-site visits. Particularly in markets created through regulation, such as water quality trading and wetland mitigation, the prospects of on-site visits by representatives of EPA or other regulatory agencies have been a deterrent to farmer participation (Breetz et al., 2004). In some markets, such as the CCX and some water quality trading programs, aggregators or other third-party service providers, rather than a government agency, verify that practices are in place. Although farmers may be less reluctant to deal with USDA-led verification for market services, such a role could put USDA at odds with its historical constituents. Experience with conservation compliance and Swampbuster (a compliance program to discourage the draining of wetlands) would seem to bear this out. The Government Accountability Office found that almost half of all NRCS field offices were not properly verifying that producers were meeting the requirements of the compliance and Swampbuster provisions (U.S. General Accounting Office, 2003). A reluctance to assume an enforcement role was cited as one of the reasons. Improved remote-sensing technology might provide more acceptable (less intrusive) means of verification, although this practice may not be applicable for all types of management options.

Verification almost always concerns management practices or land use, rather than the environmental services that are being produced. Measuring environmental services, such as water quality, carbon sequestration, wetland functions, and wildlife, is often extremely difficult and costly. Verifying practices is much less costly and is sufficient as long as market participants accept that the expected services are actually being produced.

ISSUE: COST OF INFORMATION

An important aspect of a market for environmental services is that participants have access to the information they need to make informed decisions. Producers need to know which markets they can participate in, how to produce the services demanded, and what the total cost to the farm business will be. Producers are not likely to have the time to research all the questions that need to be answered, given the time needed for managing the farm.

Government and other groups can reduce the costs of participating in a market by providing the necessary information. The USDA departmental regulation calls for USDA to conduct outreach, education, technology- transfer, and partnership-building activities with producers, using established institutional arrangements, to help producers participate in markets for environmental services. Many State cooperative extension offi ces have developed publications to help producers set up a fee-hunting business, with checklists to help identify business goals, the type of lease to offer (daily, long term, lease to a hunt club), other services to offer (bed and breakfast, guides, game cleaning), how to advertise, and how to manage risk (Chopak, 1992; Porter et al., 2007). Nongovernment organizations and private businesses that benefi t by farmer participation in markets also have an incentive to reduce producers' information costs. NutrientNet and the Nitrogen Trading Tool are examples of tools that can reduce information costs, as well as uncertainty.

Educating the public presents an important step in increasing demand for environmental services. Raising the public's awareness of the potential threats from GHG emissions could increase their willingness to pay for GHG reductions in retail markets (Trexler, Kosloff, and Silon, 2006).

ISSUE: BRINGING TOGETHER BUYERS AND SELLERS

Environmental services are produced across a diverse landscape. It may be costly for individual buyers to find all potential suppliers and to discover what each is selling, especially when the demand from a single source is much greater than the supply from a single farm. For example, a single sewage treatment plant may require nutrient credits from multiple farms to meet its permit requirements. Similarly, it can be costly for producers to find potential buyers, many of whom may be residing some distance away.

One way that markets have addressed this issue is through formal clearinghouses that assemble information from both buyers and sellers, making it easier for potential trading partners to find each other and to gauge supply and demand. The Internet is an obvious tool that could be used to facilitate trades. For example, NutrientNet, World Resources Institute's on-line nutrient-trading tool, could play a clearinghouse role in water quality trading programs (Kramer, 2003).

Government is playing a clearinghouse role in some markets. State-operated clearinghouses make it easier for point sources and nonpoint sources to find each other in some water quality trading programs (Breetz et al., 2004). The Voluntary Greenhouse Gas Reporting Registry can help agriculture and forest entities take advantage of State- and private-sector-generated opportunities to trade emission reductions and sequestered carbon.

Third-party brokers and aggregators also play a more direct role of bringing buyers and sellers together by purchasing credits from producers and selling them to buyers. Aggregators play a critical role in the Chicago Climate Exchange and are present in some water quality trading programs. In some cases, government plays an aggregator role by purchasing credits from producers and selling them on the market (such as what North Carolina does in its Tar-Pamlico trading program). State agencies serve as third-party brokers in some wetland mitigation markets to reduce uncertainty and arbitration costs. A number of State programs purchase hunting access rights from landowners and make these available to the hunting public. Hunters can consult State-provided atlases to find hunter-accessible land, with no need to seek out the individual landowner.

ISSUE: COORDINATING CONSERVATION PROGRAMS WITH MARKETS

Federally funded conservation programs and markets for environmental services can interact in several ways. USDA for the most part does not claim any credits in markets for environmental services that are created through practices implemented with financial assistance from conservation programs, allowing landowners to sell them. However, the WRP does not allow environmental services (such as carbon sequestration) created by wetland restoration to be sold. Markets for environmental services and conservation programs can also compete with each other for the same natural capital, driving up costs to the possible detriment of market development. For example, the WRP may, in some areas, reduce the stock of lands most suited to wetland restoration, leaving mitigation bankers with higher restoration costs.

Rules of individual markets may present confl icts with conservation programs. Many water quality trading programs do not allow producers to sell pollution reductions from practices financed through a conservation program, arguing that these improvements would have occurred without trading. This restriction is similar to the WRP example above. On the other hand, the Chicago

Climate Exchange has no such restriction and will pay producers for carbon sequestration from practices for which producers have already received payment (raising the question of additionality).

Coordinating conservation programs and environmental service markets can enhance the performance of both. In trading programs that establish a baseline on a minimum level of stewardship, targeting conservation programs, such as EQIP, at producers with the most serious environmental problems not only increases program performance, but could also increase the number of producers who are willing to enter a market. The policy simulation on pages 3 9-43 indicates that coordinating the CRP with fee hunting opportunities could benefi t the program as well as producers and wildlife by reducing the rental rates landowners are willing to accept while increasing their efforts to improve wildlife habitat.

Of interest is the potential impact of participation in markets for environmental services on USDA's compliance programs. Conservation compliance requires farmers to meet particular soil conservation goals in order to receive program benefi ts. Similarly, Swampbuster requires that producers not drain wetlands as a condition for receiving program benefi ts. Compliance requirements may be less costly to producers if credits produced by adopting soil- conserving practices or maintaining wetlands could be sold in water quality, carbon, or other markets.

USDA has developed a partnership agreement with EPA to coordinate agency policies and activities that promote the effective use of water quality credit trading. To this end, USDA agrees to identify and remove program barriers that might impede the development of water quality trading markets. What these are, however, will depend on the rules adopted in each market. Similar agreements could be developed for other markets as well.

ISSUE: THE ROLE OF POLICY

The design and eligibility requirements of markets for environmental services can greatly affect how attractive they are to potential participants. As discussed in the "Water Quality Markets" section of chapter 4, baseline requirements can greatly infl uence the cost and supply of credits. As shown in the greenhouse gas case study, basing credits on net sequestration rather than gross sequestration greatly affects potential returns to producers from trading.

Major expansions in some markets (i.e., wetland services, water quality, and greenhouse gases) come only with expanded or more stringent regulations on environmental quality. The low price for carbon credits in the CCX reflects the relatively low level of demand inherent in a voluntary program. A number of water quality trading programs cited lack of trades for discharge allowances because discharge caps were too high to stimulate demand (Breetz et al., 2004). Also, in a global sense, the demand for water quality improvements from producers is currently low because few impaired watersheds have opted to implement a water quality trading program and nonpoint sources are not capped.

Increased demand for environmental services from agriculture could occur when regulations change or trading programs are expanded into new areas. Requiring agricultural sources to also meet an emissions cap in a carbon market would greatly enhance demand for sequestration and result in a much larger market. Regulating all emission sources would also address the problem of leakage that occurs when payments are based on gross sequestration rather than net sequestration. Similarly, a more vigorous water quality trading market would be realized if nonpoint sources were included under a cap just as point sources are. This practice would spur nonpoint-nonpoint trading, as well as point-nonpoint.

Because of program requirements, producers considering whether to enter the wetland mitigation market face a relatively long period between starting wetland restoration and being able to sell wetland credits. A Government interested in promoting producer participation in mitigation banks could reduce these startup costs by working with lending institutions to construct loans that provide capital in increments, negotiate flexibility on loan repayment dates (perhaps delaying loan payments until wetland credits are marketed), and guarantee loans so that producers could receive a lower interest rate.

MARKETS ARE NOT ALWAYS THE ANSWER

We have shown that markets for environmental services rarely develop without some type of outside intervention. Government and other groups can reduce supply and demand impediments through regulation, market design, program coordination, education, verification, certification, and research. One of the features of working markets is the incentive to reduce transaction costs. While transaction costs may be high initially, and require Government assistance to reduce them to get the market started, costs tend to decrease over time as new institutions and mechanisms are developed by those who benefit most from them.

What the ultimate scale of markets for environmental services might be is difficult to say. For fee hunting, which is not a new concept, attitudes of both landowners and hunters may prevent much expansion. Both the water quality trading and wetland case studies indicate that the combination of factors required for markets to develop may be limited to a relatively few areas, given the current regulatory regime. On the other hand, the market for greenhouse gas reductions could be greatly expanded if a national discharge cap is implemented and producers across the country could participate in the global market. Organic agriculture and other labels are relatively new, and increased concerns over the environment could raise demand for foods produced in such a way as to provide environmental services.

Even though government can take a number of actions to promote markets for environmental services, such actions may not always be advisable. The costs of setting up and supporting a market may outweigh the benefits.

The uncertainties associated with nonpoint-source pollution from farms may never be overcome sufficiently enough to allow water quality trading markets to develop on a wide scale. Government may have to use alternative approaches, such as regulation or financial incentives, to reduce pollution from nonpoint sources and to improve water quality. Similarly, difficulties in measuring wetland services that are being lost through development or gained through restoration could relegate mitigation banking to a seldom-used tool and increase the demand for regulation or other approaches for meeting the national goal of no-net loss of wetlands. Free riding will continue to limit demand for foods covered by an eco-label, reducing the economic incentive to expand eco-friendly agriculture. Fee hunting may never become widespread because of long-ingrained attitudes about access to land for hunting.

It is probably safe to say that markets for environmental services will never supplant the need for traditional conservation programs, which will continue to play a major role in providing environmental services. Where markets do develop, government can play a role in advising market managers on the potential tradeoffs between different design and eligibility options, in providing outreach and information to reduce transactions costs and uncertainty for market participants, and in establishing standards and certification that provide consumer confidence in environmental services produced by farmers and ranchers.

REFERENCES

Altieri, M. "The Ecological Role of Biodiversity in Agroecosystems," *Agriculture, Ecosystems and Environment* 74:19-31, June 1999.

Antle, J.M. "The New Economics of Agriculture," *American Journal of Agricultural Economics* 81(5):993-1010, December 1999.

Benson, D.E. "Survey of State Programs for Habitat, Hunting, and Nongame Management on Private Lands in the United States," *Wildlife Society Bulletin* 29(1):354-58, Spring 2001a.

Benson, D.E. "Wildlife and Recreation Management on Private Lands in the United States," *Wildlife Society Bulletin* 29(1):359-71, Spring 2001b.

Benson, D.E., R. Shelton, and D.W. Steinbach. Wildlife Stewardship and Recreation on Private Lands, College Station, TX: Texas A&M University Press, 1999.

Bihrle, C. "Perceptions and Realities: Game and Fish Surveys Provide Insight into Current Issues," ND Outdoors, pp. 16-20, August 2003.

Boyd, J., and S. Banzhaf. "What are Ecosystem Services?: The Need for Standardized Environmental Accounting Units," Discussion Paper 06-02, Resources for the Future, Washington, DC, 2006.

Breetz, H.L., K. Fisher-Vander, L. Garzon, H. Jacops, K. Kroetz, and R. Terry. "Water Quality Trading and Offset Initiatives in the U.S.: A Comprehensive Survey," Hanover, NH: Dartmouth College, August 2004, Accessed at www.dartmouth.edu/~kfv/waterqualitytradingdatabase.pdf.

Bruhn, C.M., K. Diaz-Knauf, N. Feldman, J. Harwood, G. Ho, E. Ivans, L. Kubin, C. Lamp, M. Marshall, S. Osaki, G. Stanford, Y. Steinbring, I. Valdez, E. Williamson, and E. Wunderlich. "Consumer Food Safety Concerns and Interest in Pesticide-Related Information," *Journal of Food Safety* 12(3):253-62, October 1991.

Butler, M.J., A.P. Teascher, W.B. Ballard, and B.K McGee. "Commentary: Wildlife Ranching in North America—Arguments, Issues, and Perspectives," Wildlife Society Bulletin 33(1):381-89, April 2005.

Butt, T.A., and B.A. McCarl. "Farm and Forest Carbon Sequestration: Can Producers Employ it to Make Some Money?" Choices 19(3):27-31, 3rd Quarter 2004.

Cason, T., and L. Gangadharan. "Environmental Labeling and Incomplete Consumer Information in Laboratory Markets," *Journal of Environmental Economics and Management* 43(1):113-34, January 2002.

Chicago Climate Exchange. Soil Carbon Management Offsets, August 2007, Accessed at www.chicagoclimatex.com/docs/offsets/CCX_Soil_Carbon_Offsets.pdf.

Chopak, C. Promoting Fee-Fishing Operations as tourist Attractions, ID: E-2409, Michigan State University Extension, June 1992, Accessed at web1.msue.msu.edu/imp/modtd/33809023.html.

Climate Trust. Window #1: Offsets for use in the Regional Greenhouse Gas Initiative, 2007, Accessed at www.climatetrust.org/solicitations_RGGI.php.

Conner, R., A. Seidl, L. VanTasssell, and N. Wilkins. "United States Grasslands and Related Resources: An Economic and Biological Trends Assessment," Texas A&M University, Institute of Renewable Natural Resources, 2001.

Conover, M.R. "Perceptions of American Agricultural Producers about Wildlife on Their Farms and Ranches," Wildlife Society Bulletin 26(3):597-604, Autumn 1998.

Conservation Tillage Information Center. Getting Paid for Stewardship: An Agricultural Community Water Quality Trading Guide, West Lafayette, IN, 2006.

Costanza, R., R. d'Arge, R. deGroot, S. Farber, M. Grasso, B. Hannon, K. Limburg, S. Naeem, R.V. O'Neill, J. Paruelo, R.G. Raskin, P. Sutton, and M. van den Belt. "The Value of the World's Ecosystem Services and Natural Capital," *Nature* 387:253-60, May 15, 1997.

Council for Agricultural Science and Technology. Climate Change and Greenhouse Gas Mitigation: Challenges and Opportunities for Agriculture, Task Force Report No. 141, Ames, IA, 2004.

Cuperus, G., G. Owen, J.T. Criswell, and S. Henneberry. "Food Safety Perceptions and Practices: Implications for Extension," *American Entomologist* 42:201-03, Winter 1996.

Davies, A., A.J. Titterington, and C. Cochrane. "Who Buys Organic Food? A Profi le of the Purchasers of Organic Food in Northern Ireland," *British Food Journal* 97(10):17-23, 1995.

Drinkwater, L.E., P. Wagoner, and M. Sarrantonio. "Legume-Based Cropping Systems Have Reduced Carbon and Nitrogen Losses," *Nature* 396:262, November 19, 1998.

Ducks Unlimited. 2007. Accessed at www.ducks.org/Conservation/PriorityAreas/ 1599/PriorityAreasHome.html,

Ecosystem Marketplace. Backgrounder: Chicago Climate Exchange (CCX), Katoomba Group, 2007a, Accessed at ecosystemenvironmentalmarketplace. com/pages/marketwatch. backgrounder.php?market_id= 1 3&is_aggregate= 1.

Ecosystem Marketplace. Backgrounder: European Union Emissions Trading Scheme (EU ETS), Katoomba Group, 2007b, Accessed at ecosystemmarketplace.com/ pages/marketwatch.backgrounder.php?market_id= 10&is_aggregate=0.

Ecosystem Marketplace. Backgrounder: Non-Kyoto, Katoomba Group, 2007c, Accessed at ecosystemmarketplace.com/pages/marketwatch.back-grounder.php? market_id= 11&is_aggregate=0.

Ecosystem Marketplace. Marketwatch, 2007d, Accessed at ecosystem marketplace.com/pages/static/marketwatch.php.

Elkins, P. "Identifying Critical Natural Capital: Conclusions About Critical Natural Capital," *Ecological Economics* 44:277-92, March 2003.

Environmental Valuation Reference Inventory. www.evri.ca/, Accessed September 30, 2007.

Food Alliance. Accessed at foodalliance.org.

Food and Agriculture Organization of the United Nations. The State of Food and Agriculture: Paying Farmers for Environmental Services, FAO Agriculture Series No. 38, Rome, 2007.

Goldman, B., and K.L. Clancy. "A Survey of Organic Produce Purchases and Related Attitudes of Food Cooperative Shoppers," *American Journal of Alternative Agriculture* 6(2):89-92, 1991.

Greene, C. U.S. Organic Farming Emerges in the 1990s: Adoption of Certified Systems, Agriculture Information Bulletin No. 770, U.S. Department of Agriculture, Economic Research Service, June 2001.

Gross, C.M., J.A. Delgado, S.P. McKinney, H. Lal, H. Cover, and M.J. Shaffer. "Nitrogen Trading Tool to Facilitate Water Quality Credit Trading," *Journal of Soil and Water Conservation* 63(2):44A-45A, March-April 2008.

Heimlich, R., R. Claassen, K.D. Wiebe, D. Gadsby, and R.M. House. Wetlands and Agriculture: Private Interests and Public Benefits, Agricultural Economic Report No. 765, U.S. Department of Agriculture, Economic Research Service, August 1998, Accessed at. www.ers.usda. gov/publications/aer765/.

Helland, J. Walk-In Hunting Programs in Other States, Information Brief, Minnesota House of Representatives Research Department, St. Paul, MN, 2006.

Jones, W., I. Munn, S. Grado, and J. Jones. Fee-Hunting and Wildlife Management Activities by Non-industrial, Private Landowners in the Mississippi Delta, Article No. FO 123, Forest and Wildlife Research Center, Mississippi State University, 1999, Accessed at sofew.cfr.msstate.edu/papers/0114jones.pdf

Kenny, A. "Bankers, Developers and Environmentalists Weigh in on New Wetlands Regulations," in Banking on Conservation 2007: Species and Wetland Mitigation Banking, Ecosystem Marketplace, 2007.

Kieser and Associates. "Ecosystem Multiple Markets: A White Paper," Draft paper produced on behalf of The Environmental Trading Network, 2004, Accessed at www.envtn.org/docs/EMM_WHITE_PAPERApril04.pdf.

King, D.M. "Crunch Time for Water Quality Trading," Choices 20(1):7 1-76, 1st Quarter 2005.

King, D.M., and P.J. Kuch. "Will Nutrient Credit Trading Ever Work? An Assessment of Supply and Demand Problems and Institutional Obstacles," Environmental Law Reporter 33:10352-68, 2003.

Kramer, J. Lessons from the Trading Pilots: Applications for Wisconsin Water Quality Trading Policy, Resource Strategies, Inc., Madison, WI, 2003.

Lal, R., J.M. Kimble, R.F. Follett and C.V. Cole. The Potential of U.S. Cropland to Sequester Carbon and Mitigate the Greenhouse Effect, Chelsea, MI: Ann Arbor Press, 1998.

Langner, L. "Hunter Participation in Fee Access Hunting," Transactions of the 52nd North American Wildlife Natural Resources Conference, Wildlife Management Institute, pp 475-82, 1987.

Land Trust Alliance. 2005 National Land Trust Census Report, Washington, DC, November 2006, Accessed at www.lta.org/aboutus/census.shtml.

Larson, C. "The End of Hunting? How Only Progressive Government Can Save a Great American Pastime," Washington Monthly, January-February 2006.

Leopold Center. "Ecolabel Value Assessment: Consumer and Food Business Perceptions of Local Foods," Iowa State University, Ames, IA, 2003.

Lewandrowski, J., and K. Ingram. "Agricultural Resources and Environmental Indicators: Wildlife Resources Conservation," Agricultural Resources and Environmental Indicators, Agriculture Handbook No. 722, U.S. Department of Agriculture, Economic Research Service, April 2001, Accessed at www.ers.usda.gov/publications/arei/ah722/arei3_3/DBGen.htm

Lewandrowski, J., M. Peter, C. Jones, R. House, M. Sperow, M. Eve, and K. Paustian. Economics of Sequestering Carbon in the U.S. Agricultural Sector, Technical Bulletin No. 909, U.S. Department of Agriculture, Economic Research Service, April 2004.

Lipson, M. Searching for the O-word, Organic Farming Research Foundation, Santa Cruz, CA, 1997.

Lohr, L., and L. Salomonsson. "Conversion Subsidies for Organic Production: Results From Sweden and Lessons for the United States," *Agricultural Economics* 22(2): 133-46, March 2000.

Mäder, P., A. Fliebach, D. Dubois, L. Gunst, P. Fried, and U. Niggli. "Soil Fertility and Biodiversity in Organic Farming," *Science* 296(5573): 1694- 97, May 31, 2002.

Marriott, E., and M.M. Wander. "Total and Labile Soil Organic Matter in Organic and Conventional Farming Systems," *Soil Science Society of America Journa*l 70:950-59, 2006.

McCann, R.J. "Environmental Commodities Markets: 'Messy' versus 'Ideal' Worlds," Contemporary Economic Policy 14(3):85-97, July 1996.

McCarl, B.A., and U.A. Schneider. "U.S. Agriculture's Role in a Greenhouse Gas Emission Mitigation World: An Economic Perspective," Review of Agricultural Economics 22(1): 134-59, June 2000.

Mid-Atlantic Regional Water Program. A Primer on Water Quality Credit Trading in the Mid-Atlantic Region, Agricultural Research and Cooperative Extension, Pennsylvania State University, University Park, PA, 2006.

Millenium Ecosystem Assessment. Ecosystems and Human Well-being: A Framework for Assessment, Washington, DC: Island Press, 2003.

Morgan, J., B. Barbour, and C. Greene. "Expanding the Organic Produce Niche: Issues and Obstacles," Vegetables and Specialties: Situation and Outlook Report, VGS-263, U.S. Department of Agriculture, Economic Research Service, 1990.

Murtough, G., B. Aretino, and A. Matysek. Creating Markets for Ecosystem Services, Productivity Commission Staff Research Paper, AusInfo, Canberra, 2002.

Oberholtzer, L., C. Dimitri, and C. Greene. Price Premiums Hold on as U.S. Organic Produce Market Expands, VGS-308-01, U.S. Department of Agriculture, Economic Research Service, May 2005.

Organisation for Economic Co-Operation and Development. Multifunctionality in Agriculture: What Role for Private Initiatives? Paris, 2005.

Pierce, R.A. II. Lease Hunting: Opportunities for Missouri Landowners, University Extension, University of Missouri-Columbia, 1997.

Porter, M.D., R. Masters, T.G. Bidwell, and K.L. Hitch. Lease Hunting Opportunities for Oklahoma Landowners, Bulletin T-5032, Oklahoma Cooperative Extension Service, Oklahoma State University, Stillwater, OK, 2007

Ribaudo, M.O., R.D. Horan, and M.E. Smith. Economics of Water Quality Protection from Nonpoint Sources: Theory and Practice, Agricultural Economic Report No. 782, U.S. Department of Agriculture, Economic Research Service, November 1999.

Ruhl, J.B., S.E. Kraft, and C.L. Lant. The Law and Politics of Ecosystem Services, Washington, DC: Island Press, 2007.

Saam, H. Food Alliance, Personal communication, September 21, 2007.

Shabman, L., and P. Scodari. "The Future of Wetland Mitigation Banking," Choices 20(1):65-70, 1st Quarter 2005.

Shabman, L., and P. Scodari. Past, Present, and Future of Wetlands Credit Sales, Discussion Paper 04-48, Resources for the Future, Washington, DC, 2004, Accessed at www.rff.org/Documents/RFF-DP-04-48.pdf

Smolik, J.D., T.L. Dobbs, D.H. Rickerl, L.J. Wrage, G.W. Buchenau, and T.A. Machacek. Agronomic, Economic, and Ecological Relationships in Alternative (Organic), Conventional, and Reduced-Till Farming Systems, B718, Agricultural Experiment Station, South Dakota State University, September 1993.

Soil Association. The Biodiversity Benefi ts of Organic Farming, Bristol House, UK, May 2000.

Stavins, R.N. "Lessons Learned from SO2 Allowance Trading," Choices 20(1):53-58, 1st Quarter 2005.

Stavins, R.N. "Transaction Costs and Tradeable Permits," Journal of Environmental Economics and Management 29(2): 133-48, September 1995.

Sundberg, J.O. "Private Provision of a Public Good: Land Trust Membership," Land Economics 82(3):353-66, August 2006.

Tietenberg, T.H. Emissions Trading: Principles and Practice, Washington, DC: Resources for the Future, Washington, DC, 2006.

Trexler Climate + Energy Services, Inc. A Consumers' Guide to Retail Offset Providers, Prepared for Clean Air-Cool Planet, December 2006.

Trexler, M.C., L.H. Kosloff, and K. Silon. EM Market Insights: Carbon— Going Carbon Neutral: How the Retail Carbon Offsets Market Can Further Global Warming Mitigation Goals, Ecosystem Marketplace, 2006.

U.S. Department of Agriculture, Agricultural Marketing Service. "National Organic Program; Final Rule, 7 CFR Part 205," Federal Register, December 21, 2000, Accessed at www.ams.usda.gov/nop.

U.S. Department of Agriculture, Agricultural Research Service. "Gracenet: An Assessment of Soil Carbon Sequestration and Greenhouse Gas Mitigation by Agricultural Management," Project 5402-11000-008-00, December 2007, Accessed at: www.ars.usda.gov/research/projects/projects.htm?accn_no=4 11610

U.S. Department of Agriculture, Economic Research Service. "Conservation Policy: Background," April 2007a, Accessed at www.ers.usda.gov/ Briefi ng/ConservationPolicy/background.htm.

U.S. Department of Agriculture, Economic Research Service. "Conservation Policy: Land Retirement Programs," April 2007b, Accessed at www.ers.usda.gov/Briefi ng/ConservationPolicy/retirement.htm.

U.S. Department of Agriculture, Economic Research Service. "Organic Production," Accessed at www.ers.usda.gov/Data/Organic/ on October 5, 2007c.

U.S. Department of Agriculture, Economic Research Service. "Population-Interaction Zones for Agriculture (PIZA): Discussion," May 2005. Accessed at www.ers.usda.gov/Data/PopulationInteractionZones/ discussion.htm.

U.S. Department of Agriculture, Economic Research Service and National Agricultural Statistics Service. USDA Agricultural Resource Management Survey, multiple years.

U.S. Department of Agriculture, National Agricultural Statistics Service. 2002 Census of Agriculture, Vol. 1: Part 51, Chapter 2, AC-02-A-51, United States Summary and State Data, June 2004.

U.S. Department of Agriculture, Natural Resources Conservation Service. "Chief Knight Tours Nation's First Ag Wetland Mitigation Bank," NRCS This Week, April 28, 2004a, Accessed at www.nrcs.usda.gov/news/ thisweek/2004/040428/moknightwetlandbankearthday.html.

U.S. Department of Agriculture, Natural Resources Conservation Service. Conservation Effects Assessment Project (CEAP), December 2006a, Accessed at www.nrcs.usda.gov/technical/nri/ceap/ceapgeneralfact.pdf

U.S. Department of Agriculture, Natural Resources Conservation Service. eDirectives, Title 440, Pat 514 – Wetland Reserve Program, 2007a, Accessed at policy.nrcs.usda.gov/viewerFS.aspx?id=2192

U.S. Department of Agriculture, Natural Resources Conservation Service. "Electronic Field Offi ce Technical Guide," Accessed at www.nrcs.usda.gov/technical/efotg/ in October 2007b.

U.S. Department of Agriculture, Natural Resources Conservation Service. National Resources Inventory 2002 Annual NRI, 2004b, Accessed at www.nrcs.usda.gov/technical/land/nri02/nri02wetlands.html.

U.S. Department of Agriculture, Natural Resources Conservation Service. Performance Results System, 2007c, Accessed at ias.sc.egov.usda.gov/prshome/.

U.S. Department of Agriculture, Natural Resources Conservation Service. USDA Roles in Market-Based Environmental Stewardship, Departmental Regulation 56000-003, December 20, 2006b.

U.S. Department of Agriculture, Natural Resources Conservation Service, and Iowa State University, Statistical Laboratory. Summary Report: 1997 National Resources Inventory (revised December 2000), December 2000, Accessed at www.nrcs.usda.gov/technical/NRI/1997/summary_report/

U.S. Department of Agriculture, Office of the Chief Economist, Global Change Program Office. U.S. Agriculture and Forestry Greenhouse Gas Inventory: 1990-2005, 2007.

U.S. Department of Agriculture, USDA Study Team on Organic Farming. Report and Recommendations on Organic Farming, U.S. GPO No. 620-220/3641, July 1980.

U.S. Department of Energy, U.S. Energy Information Administration. Voluntary Reporting of Greenhouse Gases Program, 2007, Accessed at www.eia.doe.gov/oiaf/1605/Brochure.html.

U.S. Department of Interior, Fish and Wildlife Service. "National Wetlands Inventory: Wetland Plants," 2007, Accessed at www.fws.gov/nwi/plants.htm on August 2007.

U.S. Department of Interior, Fish and Wildlife Service, and U.S. Department of Commerce, Bureau of the Census. 2001 National Survey of Fishing, Hunting, and Wildlife-associated Recreation. 2002.

U.S. Environmental Protection Agency, AgSTAR. Market Opportunities for Biogas Recovery Systems: A Guide to Identifying Candidates for On-Farm and Centralized Systems, EPA-430-8-06-004, 2006.

U.S. Environmental Protection Agency, Office of Atmospheric Program. Inventory of U.S. Greenhouse Gas Emissions and Sinks: 1990-2004, EPA 430-R-06-002, 2006.

U.S. Environmental Protection Agency, Office of Water. America's Wetlands: Our Vital Link Between Land and Water, EPA 843-K95-001, 1995a.

U.S. Environmental Protection Agency, Office of Water. "Federal Guidance for the Establishment, Use, and Operation of Mitigation Banks," Federal Register 60(228):58605-614, November 28, 1995b, Accessed at www.epa.gov/owow/wetlands/guidance/mitbankn.html

U.S. Environmental Protection Agency, Offi ce of Water. National Section 303(d) List Fact Sheet, 2007a, Accessed at oaspub.epa.gov/waters/ national_rept. control#IMP_STATE.

U.S. Environmental Protection Agency, Offi ce of Water. National Water Quality Inventory: 2000 Report to Congress, EPA-841-R-02-001, August 2002.

U.S. Environmental Protection Agency, Pollution Prevention Division. "Environmental Labeling: Issues, Policies, and Practices Worldwide," EPA Contract No. 68-W6-0021, December 1998.

U.S. Geological Survey. SPARROW Surface Water-Quality Modeling Nutrients in Watersheds of the Conterminous United States, 2000, Accessed at water.usgs.gov/nawqa/sparrow/wrr97/results.html.

U.S. General Accounting Offi ce. Agricultural Conservation: USDA Needs to Better Ensure Protection of Highly Erodible Cropland and Wetlands, Publication No. GAO-03-418, Washington, DC, April 2003.

U.S. Government Accountability Offi ce. Wetlands Protection: Corps of Engineers Does Not Have an Effective Oversight Approach to Ensure that Compensatory Mitigation is Occurring, Report GAO-05-898, Washington, DC, September 2005,

Washington Department of Fish and Wildlife. Private Lands Wildlife Management Area (PLWMA): Private Land Partnerships for Hunter Access, Discussion paper, August 2004, Accessed at wdfw.wa.gov/wlm/plwma/plwma_accessprogram.htm on February 16, 2007.

Weaver, R.D., D.J. Evans, and A.E. Luloff. "Pesticide Use in Tomato Production: Consumer Concerns and Willingness-to-Pay," Agribusiness, 8(2):131-42, 1992.

Wiggers, E.P., and W. Rootes. "Lease Hunting: Views of the Nation's Wildlife Agencies," Transactions Of the North American Wildlife and Natural Resources Conference 52:525-29, 1987.

Wilkinson, J., and J. Thompson. 2005 Status Report on Compensatory Mitigation in the United States, Environmental Law Institute, April 2006, Accessed at www.elistore.org/reports_detail.asp?ID=11137

World Resources Institute. "About NutrientNet," Accessed at www.nutrientnet.org/about.cfm, 2007.

Wyman, J. "The Wisconsin Health Grown Potato Story," presentation, IPM Symposium, St. Louis, April 4, 2006.

Zeuli, K.A., and J.R. Skees. "Will Southern Agriculture Play a Role in a Carbon Market?" Journal of Agricultural and Applied Economics 32(2):235-48, August 2000.

APPENDIX: PREDICTING THE LOCATION OF NEW MITIGATION BANKS

If the forces that drove the demand for and, subsequently, the creation of mitigation banks continue, history can provide a perspective of where future mitigation banks might be located. Data from the USACE (edited by the Environmental Law Institute) identify counties with mitigation banks (both approved and waiting for approval). Currently, banks are dispersed across and within multiple States and in rural and urban areas, suggesting that opportunities have been widespread (see figure 4.7).

To predict the probability that a county might have a mitigation bank in the near term, a probit model is estimated using data on existing mitigation banks, county population demographics, land use, and other factors. As discussed above, one would expect that new banks will be created in counties with higher urban pressure and greater wetland acreage. Based on these and other factors, the probability model is expressed as:

Bank = f(urban pressure (low, medium, high), wetland acres, wetland acres squared, total agr land)

Where:

1. Bank equals 1 if the observation county has a mitigation bank or application and zero otherwise;
2. low (medium, high) equals 1 if urban pressures (as defined by county PIZA scores) are low (medium, high) and zero otherwise. Population-interaction zones for agriculture (PIZA) codes are derived from a classification scheme that indexes small geographic areas according to the size and proximity of population concentrations (akin to a gravity model). The

codes are discrete values ranging from 1 (rural) to 4 (high population interaction);
3. wetland acres is a county's wetland acreage;
4. wetland acres sq is wetland acres squared;
5. total agr land is the total agricultural land within the county.

Data on wetland acreage are from the National Resources Inventory (USDA, NRCS, 2000). Total agricultural land is from the 2000 Agricultural Census, and the qualitative measures of urban pressure are from ERS.

Results of the analysis are statistically significant and consistent with expectations (app. table 1). Urban pressure variables are significant determinants of the probability of a mitigation bank, and the sizes of their coefficients indicate that, as expected, the probability of a mitigation bank increases with greater urban pressure. The positive coefficient on wetland acres and negative coefficient on wetland acres sq indicate that greater wetland acreage is likely to increase the need for a mitigation bank (e.g., without wetlands, mitigation is not necessary) but at a decreasing rate as wetland acreage increases. Mitigation may be avoided if alternative lands—lands without wetlands—are available for development. The negative and significant coefficient on total agr land supports this proposition.

To provide a perspective of its reliability, we use the probit model to "back forecast"—that is, see how well the model predicts observed values. We found that the model correctly predicted counties with mitigation banks only 12 percent of the time. Counties without mitigation banks were correctly predicted 98 percent of the time. Although we have confidence that the variables in the model are appropriate, additional data related to the economics and landscape characteristics of wetland mitigation are needed to estimate a more robust prediction model.

Number of counties that have:	Number of counties predicted to have		
	Banks	No Banks	Percent correct
Banks	33	267	12%
No Banks	56	2,740	98%
Percent correct	37%	89%	90%

Note: 52 counties have no agricultural land (they are not included in the table).

Appendix table 1. Regression results: Probit model of the probability of a county's having an application for a new mitigation bank

		Parameter standard	
County variables	Estimate	Error	Pr>ChiSq
Intercept	-1.829	0.0735	<0.0001 *
Low	0.388	0.092	<0.0001*
Medium	0.801	0.104	<0.0001*
High	0.904	0.107	<0.0001*
Wetland acres	11.586	1.339	<0.0001 *
Wetland acres sq	-27.313	5.189	<0.0001*
Total agr lands	0.420	0.138	<0.0023*
Number of observations	3,101	NA	NA

Pseudo r-square = 0.06.
NA = Not applicable.
Data sources: 1. USACE district mitigation banking data as edited by the Environmental Law Institute: Dependent variable (yes/no mitigation bank). 2. USDA, NRCS, National Resources Inventory: wetland acres, and former wetlands. 3. 2000 Census of Agriculture: Net farm income, farms, and government payments, USDA, ERS (2005): Low, Medium, and High urbanization measures based on PIZA scores, www.ers.usda.gov/Data/PopulationInteractionZones/discussion.htm

ACKNOWLEDGMENTS

This report benefited from the insightful comments and information provided by Marca Weinberg, Utpal Vasavada, Kitty Smith, Mary Bohman, Pat Sullivan, Scott Swinton, Cathy Kling, Virginia Kibler, Jan Lewandrowski, Lorraine Mitchell, and the Natural Resources Conservation Service staff. Thanks also to Cynthia Nickerson for her help on the maps for the water quality trading case study, to Linda Hatcher for the excellent editorial assistance, and to Wynnice Pointer-Napper and Curtia Taylor for the design and layout.

Marc Ribaudo, LeRoy Hansen, Daniel Hellerstein, and Catherine Greene

NATIONAL AGRICULTURAL LIBRARY CATALOGING RECORD:

The use of markets to increase private investment in environmental stewardship. (Economic research report (United States. Dept. of Agriculture. Economic Research Service) ; no. 64)

1. Agricultural pollution—Economic aspects—United States.
2. Environmental protection—Economic aspects—United States.
3. Environmental policy—United States.
I Ribaudo, Marc.
II United States. Dept. of Agriculture. Economic Research Service.
III Title.

HC1 10.P55

Photo credit: Lynn Betts, USDA/NRCS.

The U.S. Department of Agriculture (USDA) prohibits discrimination in all its programs and activities on the basis of race, color, national origin, age, disability, and, where applicable, sex, marital status, familial status, parental status, religion, sexual orientation, genetic information, political beliefs, reprisal, or because all or a part of an individual's income is derived from any public assistance program. (Not all prohibited bases apply to all programs.) Persons with disabilities who require alternative means for communication of program information (Braille, large print, audiotape, etc.) should contact USDA's TARGET Center at (202) 720-2600 (voice and TDD).

To file a complaint of discrimination write to USDA, Director, Office of Civil Rights, 1400 Independence Avenue, S.W., Washington, D.C. 20250-9410 or call (800) 795-3272 (voice) or (202) 720-6382 (TDD). USDA is an equal opportunity provider and employer.

INDEX

A

abatement, 20
accounting, 10, 29, 41, 43, 76
accreditation, 63
advertising, 54, 63, 75
age, 96
agents, 63
agricultural, xi, 1, 2, 4, 7, 8, 9, 10, 11, 12, 13, 15, 19, 20, 25, 27, 31, 32, 37, 38, 39, 40, 43, 45, 49, 50, 51, 54, 55, 62, 63, 66, 68, 75, 80, 94
agricultural commodities, 1, 13, 15, 19, 54
agricultural residue, 10
agricultural sector, 32, 68
agriculture, xi, 1, 4, 7, 9, 10, 12, 18, 20, 23, 24, 25, 27, 31, 32, 34, 36, 39, 40, 41, 44, 67, 71, 73, 74, 75, 78, 80, 81, 93
aid, 20
air, vii, ix, 1, 4, 7, 8, 9, 12, 55, 72, 74
air quality, 8, 10, 72
Alabama, 48
alternative, 12, 40, 62, 66, 81, 94, 96
alternatives, 37, 56, 57, 58
ammonia, 9
amphibians, 11
AMS, 67
anaerobic, 43
anaerobic digesters, 43
Animal feed, 43

animal feeding operations, 24, 39, 43
animal waste, 8
application, 63, 93, 95
arbitration, 78
Arkansas, 48
Army, 46, 48
Army Corps of Engineers, 46, 48
ARS, 28, 30, 40
assessment, 75
assessment models, 75
assessment tools, 75
Atlantic, 28, 37, 87
atmosphere, 24, 36, 40, 41
attitudes, 53, 80, 81
authority, 46
availability, 45
average costs, 31
averaging, 52, 55
awareness, 69, 77

B

bankers, 49, 50, 78
banking, 2, 3, 4, 46, 47, 48, 49, 72, 81, 95
banks, 47, 48, 50, 51, 80, 93, 94
bargaining, xi
barrier, 20
barriers, 16, 20, 21, 79
beliefs, 96
benefits, 73

bias, 57
binding, 27, 38, 39, 72
biodiversity, 4, 11, 64, 69, 72
Biogas, 90
biomass, 10, 36
Braille, 96
Brazil, 38
breakfast, 77
breeding, 8
buffer, 9
Bureau of the Census, 52, 90
burning, 9, 10, 43
buyer, 19, 25

C

Canada, 38
capacity, x, 8, 19
caps, x, 27, 72, 80
carbon, 7, 8, 10, 19, 20, 23, 36, 37, 38, 39, 40, 41, 42, 43, 64, 71, 72, 73, 74, 75, 76, 78, 79, 80
Carbon, xii, 8, 10, 38, 40, 64, 72, 84, 85, 86, 87, 88, 89, 91
carbon credits, 38, 39, 40, 79
carbon dioxide, 37, 38
carbon emissions, 20, 38, 42
case study, 51, 79, 95
cation, xi, 39, 56, 63, 68, 69, 72, 75, 76, 80, 81, 93
Census, 36, 52, 86, 89, 90, 94, 95
centralized, 19
certification, 75
CH4, 10
chemicals, 8, 12, 65, 75
citizens, 39, 52
Civil Rights, 96
Clean Air Act, 2, 12
Clean Water Act, 3, 9, 12, 23, 27, 29, 46, 71
cleaning, 77
climate change, 39
CO2, 37, 38, 41
coal, 38
coal mine, 38
Cochrane, 66, 84

codes, 47, 93
collaboration, 10, 52
College Station, 83
Colorado, 48, 57
Columbia, 87
combustion, 9, 43
commodity, 1, 17, 19, 25, 39, 41, 58
communication, 57, 88, 96
communities, 50
community, 31
competition, x, 4
competitive advantage, 51
competitive markets, 15
complement, vii
compliance, 2, 3, 46, 49, 76, 79
complications, 55
composition, 39
concentration, 65
confusion, 66, 69
Congress, 63, 91
conservation, vii, ix, x, xi, 2, 4, 8, 9, 10, 12, 19, 20, 30, 31, 35, 40, 41, 42, 46, 52, 55, 62, 72, 73, 74, 75, 76, 78, 79, 81
Conservation Security Program, 3, 42
construction, 47, 49
consumers, 4, 7, 8, 11, 15, 17, 39, 46, 62, 63, 66, 68, 69, 75
consumption, 17, 18
contamination, 63
contracts, 18, 35, 38, 54, 55
control, vii, ix, xii, 8, 11, 24, 25, 28, 37, 46, 52, 54, 65, 91
conversion, 11
corn, 41
corporations, 38
costs, x, xi, xii, 4, 16, 18, 21, 24, 26, 28, 29, 30, 31, 41, 43, 49, 50, 54, 55, 62, 65, 68, 72, 77, 78, 80, 81
cost-sharing, 58
coverage, 43, 57, 72
covering, 38
credit, 4, 25, 32, 34, 35, 40, 42, 43, 72, 79, 96
credit market, 32, 43
critical habitat, 11
crop production, 40

Index

crop rotations, 10, 40, 65
croplands, 10
crops, 7, 40, 53, 54, 63, 64
CRP, 2, 3, 55, 56, 57, 58, 60, 61, 62, 79
CTA, 2, 3
cultivation, 8, 10
customers, 19, 39

D

dairy, 43
Dartmouth College, 83
decisions, 1, 7, 17, 18, 53, 54, 56, 76
delivery, 29
demand curve, 17
demographics, 93
Department of Agriculture, vii, ix, 4, 85, 86, 87, 88, 89, 90, 96
Department of Commerce, 52, 90
Department of Energy, 41, 90
Department of Interior, 52, 90
destruction, 38, 40
disability, 96
discharges, x, 24, 26, 27, 28, 32, 35
disclosure, 39
discrimination, 96
displacement, 38
disputes, 72
distribution, 19, 34, 47, 54, 56, 57, 60, 68
diversity, 8, 10
dominance, 10, 52
drainage, 47
drought, 1
dust, 2

E

ears, 41
ecological, 11
Ecological Economics, 85
economic incentives, 44
economic losses, 49
economic performance, 75

Economic Research Service, 12, 64, 85, 86, 87, 88, 89, 96
economics, 94
ecosystem, 7, 16, 85
ecosystems, 7, 11, 44
education, 2, 3, 75
electricity, 43
emission, 8, 28, 37, 38, 41, 42, 43, 78, 80
emission source, 38, 80
Endangered Species Act, 3, 12
energy, 39, 43, 64, 68
Energy Information Administration, 90
Energy Policy Act, 41
engines, 9
enrollment, 56
enterprise, 54
environment, ix, xii, 24, 44, 66, 81
environmental impact, 68, 73
environmental policy, 2
environmental protection, 63
Environmental Protection Agency, 4, 33, 34, 35, 41, 90, 91
Environmental Quality Incentives Program, 3, 12, 30, 42
EPA, 4, 9, 10, 11, 25, 30, 32, 43, 46, 66, 71, 73, 76, 79, 90, 91
EQIP, 2, 3, 30, 31, 79
erosion, 11, 40
estimating, 40, 41, 49, 74
estuaries, 9
estuarine, 9
ethane, 38
European Union (EU), 37, 38, 85
exercise, 10, 52
expansions, 79
externalities, 7, 19

F

failure, 12, 15
familial, 96
FAO, 85
farmers, vii, ix, x, xi, xii, 1, 4, 19, 30, 31, 41, 42, 43, 49, 51, 52, 53, 54, 58, 64, 66, 68, 74, 76, 79, 81

Index

farming, 19, 55, 64, 65, 66, 68, 73, 74, 76
farmland, 7, 12, 49, 51, 64, 68
farms, x, 7, 8, 10, 15, 21, 23, 29, 30, 31, 34, 35, 36, 41, 51, 52, 53, 65, 69, 71, 74, 77, 81, 95
fear, 29
February, 41, 86, 91
Federal Insecticide, Fungicide, and Rodenticide Act, 3, 12
Federal Register, 88, 90
Federal Trade Commission, 67
fee, 12, 39, 50, 51, 52, 53, 54, 55, 56, 61, 62, 74, 77, 79, 80
feeding, 11
fencing, 9, 54
fermentation, 10
fertility, 63
fertilization, 54
fertilizer, 10, 39
fiber, ix, 7, 72
FIFRA, 3
finance, 37
financial resources, 23, 31
financial support, 44
financing, 37
fines, 24
firms, x, 17, 24, 31, 72
fish, 11, 46
Fish and Wildlife Service, 46, 50, 52, 90
fishing, 8, 46, 53
flexibility, 24, 80
flight, 75
flood, vii, ix, 8, 11, 74
flow, 4, 8, 41
food, vii, ix, 7, 8, 63, 64, 66, 67, 68, 71, 72
food safety, 63, 66
forestry, 38
Forestry, 90
forests, ix, 7, 10, 40, 41, 42, 46
Fox, 27
free riders, 16
freshwater, 11
fruits, 65
FTC, 67
fuel, vii, ix, 7

funding, 36, 39
funds, 46

G

gas, 10, 36, 41, 72
gases, 10, 36
gauge, 45, 56, 77
genetic information, 96
Georgia, 48
GHG, 10, 36, 37, 38, 39, 40, 41, 42, 43, 77
global climate change, 10, 36
global warming, 10, 38, 40, 64
Global Warming, 88
goals, 4, 24, 27, 50, 63, 77, 79
goods and services, 4, 15, 17, 23, 76
government, vii, xi, xii, 1, 4, 5, 10, 12, 15, 20, 21, 23, 29, 44, 50, 52, 58, 62, 63, 66, 68, 71, 73, 75, 76, 78, 81, 95
government intervention, 4, 5
government policy, 62
GPO, 90
grain, 64
grasses, 40, 54
grassland, 8, 11
grasslands, ix, 10, 11
gravity, 93
grazing, 10, 41, 54
Great Lakes, 9
greenhouse, x, 8, 10, 19, 20, 36, 37, 38, 40, 41, 71, 72, 79, 81
Greenhouse, 10, 36, 37, 41, 72, 78, 84, 86, 87, 89, 90
greenhouse gas, x, 8, 10, 19, 20, 36, 37, 38, 40, 41, 71, 72, 79, 81
greenhouse gas (GHG), 10
greenhouse gases, x, 10, 19, 36, 38, 40, 79
greenhouse gases (GHG), 36
groundwater, 66
groups, 4, 38, 76, 77, 80
growth, 39
guidance, 41, 90
guidelines, 73
guns, 62

H

habitat, vii, ix, 3, 7, 10, 11, 12, 51, 52, 53, 54, 55, 57, 61, 62, 68, 71, 72, 74, 79
handling, 54, 63, 65
harmful effects, 46
harvest, 54
hazards, 54
health, 9
heavy metal, 63
heavy metals, 63
heterogeneity, 59
host, vii, ix, 8
House, 85, 86, 87, 88
humans, 64
hunting, xii, 8, 46, 51, 52, 53, 54, 55, 56, 57, 58, 60, 61, 62, 71, 72, 74, 77, 78, 79, 80, 81
hydrogen, 9

I

Idaho, 48
Illinois, 25, 48
implementation, 28
incentive, xi, 9, 15, 16, 23, 42, 51, 56, 66, 77, 80, 81
incentives, 12, 21, 42, 54, 81
income, ix, 13, 15, 17, 36, 41, 43, 49, 51, 52, 53, 54, 58, 60, 61, 71, 95, 96
incomes, 1
Indiana, 48
indication, 12
indicators, 58
industrial, 86
industry, xi, 39
infrastructure, 65, 69
initiation, 63
innovation, 4
insight, xii, 5
inspection, 29
inspections, 46
institutions, 15, 39, 80
instruments, 2, 75
insurance, 75

intangible, 39
interaction, 4, 17, 39, 93
interactions, 72
internal reduction, 38
Internet, 77
intervention, 4, 5, 80
investment, vii, 20, 30, 49, 96
irrigation, 9, 40

J

justice, 66

K

Kentucky, 48
King, 29, 86
Kyoto Protocol, 37

L

labeling, vii, xii, 2, 3, 62, 63, 66, 68, 69, 71, 72
labor, 65
Lafayette, 84
lakes, 9
land, x, xii, 1, 4, 7, 8, 10, 11, 12, 15, 30, 39, 41, 42, 45, 49, 50, 51, 52, 53, 54, 55, 56, 57, 58, 59, 60, 61, 62, 72, 73, 76, 78, 81, 89, 93, 94
land use, 1, 7, 11, 30, 39, 42, 49, 76, 93
landfill, 38
land-use, 15, 41
law, 23
lawyers, 18
leadership, 39
leakage, 42, 80
legislation, 49
lending, 80
liability insurance, 55
likelihood, 51, 59, 61
limitations, x, 37
liquids, 43
livestock, 9, 43

Livestock, 36
loans, 80
local government, 1, 44, 46, 73
location, 19, 27, 28, 37, 56
long period, 80
losses, 44
Louisiana, 48

M

Madison, 86
Maine, 64
maintenance, 8, 43, 46
mammals, 11
management, x, xi, 1, 7, 8, 9, 10, 12, 17, 19, 20, 24, 28, 30, 32, 34, 35, 36, 39, 40, 42, 43, 46, 52, 54, 62, 63, 64, 73, 74, 75, 76
management practices, x, xi, 9, 10, 12, 19, 24, 30, 32, 39, 40, 42, 76
manure, 9, 10, 40, 43
marital status, 96
market, vii, ix, x, xi, xii, 1, 3, 4, 11, 13, 15, 17, 18, 19, 20, 21, 23, 24, 25, 27, 29, 30, 31, 32, 34, 36, 37, 38, 39, 41, 42, 43, 49, 50, 51, 53, 55, 56, 65, 72, 73, 74, 75, 76, 77, 78, 79, 80, 81, 85
market failure, 15
market supply curve, 17
marketing, 46, 53, 62, 63, 65, 69
marketplace, 85
markets, vii, ix, x, xi, xii, 1, 4, 8, 12, 13, 15, 16, 17, 18, 19, 20, 21, 23, 25, 29, 31, 34, 36, 37, 38, 39, 40, 41, 42, 43, 44, 46, 49, 50, 52, 54, 62, 64, 71, 72, 73, 74, 75, 76, 77, 78, 79, 80, 81, 96
marshes, 11
Maryland, 48
Massachusetts, 27
measurement, 39, 40
measures, 3, 51, 56, 94, 95
median, 60
methane, 8, 9, 10, 38, 39, 40, 43
metric, 10, 38, 43
Mexico, 38
Miami, 27

migration, 11
Millennium, 7
Minnesota, 25, 27, 64, 86
missions, 10, 24, 37, 41, 85
Mississippi, 48, 51, 52, 54, 60, 86
Missouri, 48, 87
modeling, 74
models, 28, 72, 74
motivation, 5
movement, 75

N

National Marine Fisheries Service, 46
national origin, 96
national product, 63
natural, vii, ix, x, 7, 8, 12, 15, 78
natural capital, x, 8, 12, 78
natural resources, vii, ix
Natural Resources Conservation Service, xi, 4, 8, 9, 20, 34, 35, 44, 74, 89, 90, 95
Nebraska, 48, 57, 58
negotiating, 50
negotiation, 31
nesting, 7, 11
net price, 16
New Jersey, 48
New York, 48, 64, 68
NGO, 49
NGOs, 50
nitrogen, 9, 20, 29, 30, 32, 33, 34
nitrogen gas, 9
nitrogen oxides, 9
nitrous oxide, 10
nongovernmental, 37, 49
nongovernmental organization, 37, 49
North America, 84, 86, 91
North Carolina, 48, 78
Northeast, 61, 65
Northern Ireland, 84
nutrient, 7, 8, 9, 19, 32, 34, 35, 40, 63, 64, 74, 77
nutrients, 3, 9, 32, 63
nutrition, 63

O

obligations, 24
observations, 48, 57, 95
odors, 9
OFPA, 63
Ohio, 48
oilseed, 64
Oklahoma, 48, 88
online, 30, 55, 77
open space, 11, 12, 20, 62
open spaces, 11
Oregon, 37, 48, 64
organic, xi, 8, 9, 10, 40, 63, 64, 65, 66, 67, 68, 69, 74, 75
organic compounds, 9
Organic Foods Production Act, 63
organic matter, 8, 10, 63, 64
organization, 37
organizations, 29, 45, 49, 66, 77
outreach programs, 31
overload, 72
ownership, 10, 16, 52
oxide, 10

P

paper, 83, 86, 87, 88, 91
Paris, 87
partnership, 73, 77, 79
pasture, 8, 55, 65
pathogenic, 63
pathogens, 9
penalties, 24, 63
Pennsylvania, 64, 87
perception, 12
perceptions, 53
performance, ix, xi, 19, 28, 30, 72, 73, 74, 75, 76, 79
permit, 24, 56, 71, 77
personal, 16, 17, 53, 55, 66
personal values, 16
pest control, 65
pesticide, 9, 63, 64, 66
pesticides, 3, 9, 68
phosphorus, 25, 27, 32, 33, 34
planning, 42
plants, 8, 23, 90
play, xi, xii, 10, 11, 31, 47, 74, 77, 78, 81
policy initiative, 69
policy instruments, 4
pollutant, x, 24, 25, 28, 31
pollutants, 9, 24, 29
pollution, x, xii, 7, 9, 11, 20, 23, 24, 25, 26, 27, 31, 32, 35, 64, 71, 74, 78, 81, 96
population, 1, 53, 57, 93
positive externalities, 7
potato, 68
power, 24, 37
power plant, 24, 37
power plants, 24, 37
predators, 11
prediction, 94
premium, 66, 69
premiums, 69, 75
President Bush, 41
pressure, 47, 51, 56, 58, 59, 60, 93, 94
prices, 1, 17, 18, 19, 23, 39, 41, 50, 58, 66
private, vii, x, 8, 10, 11, 12, 16, 24, 25, 31, 38, 39, 41, 46, 51, 52, 54, 55, 57, 62, 63, 66, 68, 74, 75, 77, 78, 96
private good, x, 16, 24, 25, 46, 51, 62
private investment, vii, 96
private property, 52
private-sector, 41, 63, 78
probability, 51, 93, 94, 95
producers, xi, 1, 4, 8, 9, 12, 13, 15, 17, 19, 20, 24, 29, 30, 31, 35, 36, 43, 46, 49, 50, 51, 53, 55, 66, 71, 73, 74, 75, 76, 77, 78, 79, 80, 81
product market, 17
production, ix, x, xi, 1, 4, 7, 9, 12, 17, 18, 19, 24, 25, 28, 42, 43, 54, 55, 63, 64, 65, 66, 68, 69
production possibility frontier, 17
production technology, 24
productivity, 2, 64

program, xi, 3, 20, 24, 25, 27, 29, 30, 31, 37, 38, 41, 42, 45, 52, 54, 55, 56, 57, 58, 59, 61, 63, 68, 71, 76, 78, 79, 80, 96
proliferation, 54, 66, 69
promote, 4, 21, 52, 54, 56, 79, 81
property, 37, 54, 62, 72
proposition, 94
protection, 11, 36, 50, 63, 68, 96
protocols, 39
public, vii, ix, x, xi, 12, 16, 18, 21, 24, 28, 44, 45, 46, 50, 51, 52, 55, 56, 64, 66, 69, 71, 77, 78, 96
public goods, 16, 24, 44, 66, 71
public health, 64
public investment, vii
pumps, 9

Q

quality improvement, 9, 24, 30, 80
quality of service, x, 19, 44

R

race, 96
range, 1, 7, 8, 17, 28, 41, 45, 55, 58
rangeland, 11, 38, 40
rating agencies, 39
recognition, 39
recovery, 43, 49
recreation, 11, 51, 52, 53, 54, 57
recreational, 11, 55
reduction, 8, 20, 26, 28, 29, 35, 36, 38, 41, 42, 52, 58, 60, 62, 74
regional, 37, 40, 60, 68, 74
Registry, 41, 72, 78
regulation, x, xii, 1, 8, 27, 29, 46, 73, 76, 77, 80, 81
regulations, 9, 12, 24, 44, 50, 63, 69, 79, 80
regulators, 18
reliability, 94
religion, 96
renewable energy, 38, 39

research, xi, 28, 39, 40, 66, 69, 73, 74, 75, 77, 80, 89, 96
reservation, x
residues, 63, 64
resources, vii, ix, 1, 8, 9, 10, 15, 17, 18, 20, 23, 31
retail, 39, 71, 75, 77
retirement, 2, 3, 42, 55, 89
returns, ix, x, 17, 19, 29, 30, 41, 42, 53, 69, 79
revenue, 32, 53
rice, 10
risk, 20, 49, 51, 75, 77
risks, 38, 65, 68
rivers, 9
Rome, 85
rotations, 63
runoff, 7, 8, 20, 25, 29, 75
rural, 53, 62, 93, 94

S

safety, 66
sales, 36
salmon, 68
salt, 11
salts, 9
sampling, 74
savings, 25, 26
scientists, 64
scores, 58, 59, 93, 95
search, xi
Secretary of Agriculture, 73
sediment, 8, 9, 74
sedimentation, 2
seller, 16
sensing, 76
service provider, 76
services, vii, ix, x, xi, xii, 1, 4, 7, 8, 11, 12, 15, 16, 17, 18, 19, 20, 21, 23, 31, 44, 45, 46, 47, 49, 50, 51, 54, 55, 56, 62, 63, 66, 69, 71, 72, 73, 74, 75, 76, 77, 78, 79, 80, 81
sewage, 23, 77
sex, 96
sexual orientation, 96
shape, 17

Index

shareholders, 39
sharing, 2, 3, 58
signals, vii, ix, 1
simulation, 28, 56, 79
skills, 17, 39
SO2, 88
social justice, 66
social welfare, 17
soil, 8, 9, 10, 11, 36, 38, 40, 49, 55, 63, 64, 74, 79
soil erosion, 55, 63
soils, 8, 10, 20, 37, 39, 40, 51, 64
South Carolina, 48
South Dakota, 48, 57, 58, 88
soybean, 64
spatial, 34
specialty crop, 65
species, 8, 11, 12, 51, 64, 71
speculation, 37
sponsor, 46
St. Louis, 91
standards, xi, 39, 40, 63, 66, 68, 69, 72, 75, 76, 81
Standards, 63, 72, 75
stock, 41, 78
storage, 8, 9, 10, 43, 74
strategies, 1, 36, 64
streams, 51
substances, 63
substitutes, 65
sulfur, 24
sulfur dioxide, 24
summer, 40
suppliers, 4, 49, 55, 77
supply, vii, 4, 9, 11, 12, 17, 19, 20, 23, 25, 29, 30, 31, 32, 34, 40, 42, 44, 49, 50, 51, 53, 55, 69, 74, 77, 79, 80
sustainability, 74
swamps, 11
Sweden, 87
systems, 9, 19, 40, 43, 63, 64, 65, 66, 68

T

targets, 37, 38

taxpayers, 20
technical assistance, 1, 2, 3
technology, 17, 24, 76, 77
Tennessee, 48
tenure, 42
Texas, 48, 83, 84
threatened, 11
threats, 77
time, 18, 20, 24, 46, 49, 50, 52, 54, 68, 77, 80, 94
time consuming, 50
timing, 8, 31
title, 12
TNC, 45
tolerance, 53
topology, 49
tourist, 84
toxic, 8, 68
trade, xi, 19, 25, 28, 29, 37, 38, 41, 78
trading, vii, xii, 4, 20, 23, 24, 25, 26, 27, 28, 29, 30, 31, 34, 35, 36, 38, 39, 40, 42, 46, 72, 73, 74, 75, 76, 77, 78, 79, 80, 81, 95
trading partners, 29, 31, 77
training, 54
transaction costs, 4, 16, 18, 29, 31, 41, 50, 55, 80
transactions, 18, 41, 46, 72, 81
transfer, 77
transition, 65
transparency, 41
transport, 65
transportation, 68
trees, 8, 54
tribal, 46
trusts, 4, 12

U

U.S. Department of Agriculture, ix, 4, 85, 86, 87, 88, 89, 90, 96
U.S. Department of Agriculture (USDA), ix, 4, 96
U.S. Geological Survey, 32, 33, 34, 35, 91
UK, 88

Index

uncertainty, xi, 16, 19, 20, 21, 28, 30, 39, 40, 41, 42, 69, 72, 73, 74, 75, 77, 78, 81
uniform, 63
unit cost, xi
United Nations, 7, 8, 85
United States, vii, 8, 10, 11, 12, 25, 32, 36, 37, 38, 41, 43, 51, 52, 63, 68, 83, 84, 87, 89, 91, 96
urban areas, 47, 93
urbanization, 62, 95
urbanized, 53, 57
USDA, ix, xi, xii, 4, 8, 9, 10, 11, 12, 20, 31, 32, 33, 34, 35, 36, 40, 42, 43, 44, 45, 47, 48, 49, 50, 52, 53, 56, 59, 60, 61, 62, 63, 64, 65, 66, 67, 71, 73, 74, 75, 76, 77, 78, 79, 89, 90, 91, 94, 95, 96
Utah, 48

V

Valdez, 83
validation, xi
values, ix, 1, 16, 17, 23, 62, 94
variable, 95
variables, 94, 95
vegetables, 65
vegetation, 15
Vermont, 64
visible, 40
vision, 39, 57
voice, 96

W

water, vii, ix, x, xii, 1, 4, 7, 8, 9, 10, 11, 12, 16, 19, 20, 23, 25, 27, 28, 29, 30, 31, 32, 36, 37, 40, 42, 46, 55, 62, 63, 64, 68, 69, 72, 73, 74, 76, 77, 78, 79, 80, 81, 91, 95
water quality, x, xii, 4, 7, 8, 9, 16, 19, 20, 23, 25, 27, 28, 29, 30, 31, 32, 36, 37, 42, 46, 55, 62, 68, 69, 72, 73, 74, 76, 77, 78, 79, 80, 81, 95
water resources, 9
water table, 8
waterfowl, 45, 54
watershed, 24, 29, 30, 47, 68, 73, 75
watersheds, 27, 31, 32, 34, 35, 36, 80
waterways, 24, 55
Weinberg, 95
wellbeing, 17
wells, 54
wetland restoration, 20, 49, 50, 78, 80
wetlands, ix, x, 8, 9, 10, 11, 19, 40, 44, 45, 46, 47, 49, 51, 74, 76, 79, 81, 90, 94, 95
wildlife, vii, ix, x, 1, 4, 7, 8, 10, 11, 23, 46, 51, 52, 53, 54, 55, 56, 57, 58, 59, 60, 61, 62, 63, 66, 71, 72, 74, 75, 76, 79
Wildlife Habitat Incentives Program, 3
wind, 11
Wisconsin, 25, 48, 64, 68, 86, 91
woodland, 8
workers, 66
World Resources Institute, 30, 77, 91

Y

yes/no, 95
yield, 61